Springer Theses

Recognizing Outstanding Ph.D. Research

Aims and Scope

The series "Springer Theses" brings together a selection of the very best Ph.D. theses from around the world and across the physical sciences. Each volume has been selected for its scientific excellence by an internationally renowned institute or department. For greater accessibility to non-specialists, the published versions include an extended introduction, as well as a foreword by the student's supervisor explaining the special relevance of the work for the field. The series as a whole provides a valuable resource both for newcomers to the research fields, and for other scientists seeking detailed background information on special questions. Finally, it serves as an accredited documentation of the valuable contributions made by today's younger generation of scientists.

How to Apply

Theses prepared at internationally leading universities and institutes may be nominated for publication by supervisors with the endorsement of the head of department. Each such department is entitled to nominate one thesis per year. Enquiries and requests for a nomination form should be sent, together with the thesis, to Ms. Nathalie Berg (see link on right).

Eligibility Criteria

To qualify for consideration, theses should fulfil the following criteria:

- They must be written in good English.
- The topic should fall within the confines of Chemistry, Physics, Earth Sciences, Engineering and related interdisciplinary fields such as Materials, Nanoscience, Chemical Engineering, Complex Systems and Biophysics.
- The work reported in the thesis must represent a significant scientific advance.
- A maximum of 30% of the text may be a verbatim reproduction from the author's previous publications.
- They must have been examined and passed during the 12 months prior to nomination.
- The theses should have a clearly defined structure including an introduction accessible to new PhD students and scientists not expert in the relevant field.
- After acceptance, theses should be supplemented by a "Supervisor's Foreword" outlining the wider scientific relevance of the work

Indexed by zbMATH.

Ana Cisnal de la Rica

Development of Biocooperative Control Strategies for Neuromotor Rehabilitation Robotic Platforms

A Real-Time Embedded Approach for Enhanced Human-Robot Interaction

Doctoral Thesis accepted by
University of Valladolid, Valladolid, Spain

Springer

Author
Dr. Ana Cisnal de la Rica
Medical Robotics Research Group
Institute of Advanced Production Technologies
University of Valladolid
Valladolid, Spain

Supervisors
Dr. Juan Carlos Fraile Marinero
Medical Robotics Research Group
Institute of Advanced Production Technologies
University of Valladolid
Valladolid, Spain

Dr. Javier Pérez Turiel
Medical Robotics Research Group
Institute of Advanced Production Technologies
University of Valladolid
Valladolid, Spain

ISSN 2190-5053 ISSN 2190-5061 (electronic)
Springer Theses
ISBN 978-3-032-02484-8 ISBN 978-3-032-02485-5 (eBook)
https://doi.org/10.1007/978-3-032-02485-5

This Springer imprint is published by the registered company Springer Nature Switzerland AG
The registered company address is: Gewerbestrasse 11, 6330 Cham, Switzerland

To all who have stood by me
throughout these years

Supervisors' Foreword

It is our honor to introduce the thesis work of Dr. Ana Cisnal de la Rica, who carried out a thorough and outstanding research that lead to major contributions in the framework of biomedical engineering and, particularly, in the context of upper-limb rehabilitation robotics. The path that leads to the thesis began in 2016, when Dr. Cisnal started her collaboration with our research group (Medical Robotics—Institute of Advanced Production Technologies) at University of Valladolid, Spain. Her first contact with biomedical signal processing was the study that Dr. Cisnal carried out during her final degree project of the bachelor's degree in Industrial Electronics and Automation Engineering ("Design and development of a measuring instrument for the characterization of electrodes in the context of functional electrostimulation", 2017). She then expanded her expertise in robotics through her master's thesis in Industrial Engineering ("Sensorization and control of an exoskeleton for hand rehabilitation", 2019). Since then, she was endeavoring to gain deeper knowledge on biomedical signal processing, embedded systems and robotics. As a natural consequence, she embarked on her career as Ph.D. student in the Industrial Engineering Doctoral Programme. In the period 2019–2023, Dr. Cisnal collaborated in several R&D projects, working in electronic acquisition systems and embedded systems design, applied to rehabilitation and surgical robots. She defended her doctoral thesis at University of Valladolid in September 2023 (*Development of Biocooperative Control Strategies for Neuromotor Rehabilitation Robotic Platforms*), in which she obtained the highest qualification (Summa Cum Laude).

The thesis of Dr. Cisnal presents original research focused on designing affordable, real-time solutions for physiological data acquisition and biocooperative control strategies for individuals with neuromotor impairments. Unlike traditional rehabilitation robotics, which focuses on biomechanical data, this work incorporates physiological measurements to engage the patient in the feedback loop. The studies primarily address upper-limb motor rehabilitation, which is crucial for stroke survivors who experience upper-limb paresis, impacting their quality of life. The thesis designs accessible EMG systems and wearable multimodal platforms to facilitate upper-limb biocooperative control in clinical settings. It also implements biocooperative control strategies, including EMG-based control with virtual reality and adaptive assistive control for wrist rehabilitation. These low-cost systems were validated, demonstrating high versatility and signal quality. Overall, this research offers new insights into designing affordable, effective robotic devices for upper-limb neurorehabilitation and enhances human-robot interaction in neuromotor rehabilitation.

Valladolid, Spain
July 2025

Dr. Juan Carlos Fraile Marinero
Dr. Javier Pérez Turiel

Foreword by the Bioengineering Group of the Automatic Control Spanish Association (CEA)

We humans do not notice the extraordinary complexity behind any movement we make with our limbs. It is an extraordinarily complex task requiring synergistic integration of the musculoskeletal system and the entire nervous system. Most neurological impairments impact the capacity of limb movements, impairing our ability to execute precise and energy-efficient movements. The work presented in this book, written by Dr. Ana Cisnal, contributes to the area of bioengineering and rehabilitation engineering, providing significant improvements in the use of electromyography (EMG) signals for biofeedback and control of upper limb rehabilitation robots.

Dr. Cisnal addresses critical challenges in the clinical adoption of upper limb robotic neurorehabilitation devices for patients, such as performance, reliability, and cost. The common thread of the book is how to improve and efficiently manage the EMG signal to overcome these challenges. Two robots -hand and wrist robots- and three studies support the exploration of different modalities of EMG-driven control: firstly, by developing a non-pattern recognition-based EMG-driven control in real-time, which demonstrated a notable performance in terms of hand gesture recognition accuracy and time response. Then, the platform was improved by including EMG-based visual feedback, which enabled the users to gain control on the motion of the EMG-driven exoskeleton thanks to visually observing their muscle activations. Lastly, the platform was improved by integrating other physiological measures conveying information of user involvement, performance, and intention (movement -inertial measurement unit-, electrocardiogram, galvanic skin response, and skin thermometer) within a cost-effective, wearable, embedded multimodal platform. The performance of this multimodal acquisition platform is validated through the implementation of two bio-cooperative control strategies: an EMG and IMU-based control using virtual reality-based therapy and an adaptive assistive control.

The book is based on the first author's doctoral thesis, awarded for best Ph.D. thesis in bioengineering during the 2024 edition of the award call organized by the Bioengineering Group of the Spanish Committee of Automatic Control (Comité Español de Automática, CEA). This annual award is aimed at recognizing the outstanding Ph.D. research in the field of bioengineering related to control engineering and robotics.

Participation requires at least one of the thesis' supervisors to be a partner of CEA and a member of the Bioengineering Group. The jury is composed of three well-known doctors in the field, two of them partners of CEA. Ana Cisnal's Ph.D. thesis was awarded by the Bioengineering Group of CEA as the best among other excellent candidates. Through this publication, we hope that Ana's work reaches a large international audience and becomes a valuable source of information and inspiration for other researchers and students in this field, which holds tremendous promise for advancing human health and well-being. We do not doubt that Ana's work will continue to inspire and shape the field of bioengineering, and I am confident that his contributions will have a lasting impact on the scientific community and society as a whole. Once again, congratulations to Ana on this exceptional achievement, and all the best in your future endeavors.

Dr. Antonio J. del-Ama
Bioengineering Systems
and Technologies Research Group
Rey Juan Carlos University
Móstoles, Spain

Acknowledgments First of all, I would like to express my sincere gratitude to the supervisors of this doctoral thesis, Dr. Javier Pérez Turiel and Dr. Juan Carlos Fraile Marinero, for their trust and support throughout all these years. I still remember back in 2016 when Juan Carlos proposed the possibility of doing a research stay abroad to his fourth-year students. I believe that, at that moment, my research profile began to take shape. Thank you, Juan Carlos, for giving me the opportunity to go to IBMT, to do an internship at CARTIF, and for offering me the chance to start working at ITAP Medical Robotics. Thank you, Javier, for the rigor you always try to bring to my research work, your valuable suggestions, and the careful reviews you have done on the papers that make up this thesis. Thank you both for the unconditional support you have always given me since I began working with you.

I would also like to thank all the colleagues who are or were part of the ITAP Medical Robotics group. With special affection, I want to thank those who were there at the beginning: Víctor and David, without whom this research would not have been possible. But also, to those who joined later: Guillermo, Gonzalo, Alberto, Wiki, Diego, Daniel, Carlos... Thank you for all the personal and professional support, and for all the good times we've shared.

I would also like to express my sincere gratitude to Dr. Robert Riner and Dr. Diego Páez-Granados for welcoming me into their research group. Thank you, Diego, for your supervision, suggestions, and dedication. Furthermore, I would like to acknowledge and thank all the members of the SCAI Lab, with special thanks to Mehdi, Sabrina, Yankee, Siri, and Bertram. They not only contributed professionally to my research work but also personally, providing me with many memorable moments. Please know that you are always welcome to visit your second home in Valladolid.

However, a research project is also the result of the recognition and unconditional support we receive from those who love us, without which we would not have the energy that motivates us to improve both as people and as professionals. Thank you to my parents and friends, who have always given me great moral and emotional support. But especially, thank you, Rafael, for always being there, for your patience and understanding.

To all of them, thank you for your friendship, support, and contributions throughout all these years. Without you, this work would never have been written, and that is why this work is also yours. To all, thank you very much.

Finally, I gratefully acknowledge the financial support that made this research possible. This work was in part supported by the Regional Ministry of Education grant for the pre-doctoral recruitment of research staff (ORDEN de 12 de Diciembre de 2019, de la Consejería de Educación de la Junta de Castilla y León), co-financed by the European Social Fund (ESF).

Competing Interests The author has no competing interests to declare that are relevant to the content of this manuscript.

Part of this thesis has been published in the following articles

Journals

1. **Ana Cisnal**, Javier Pérez-Turiel, Juan Carlos Fraile, David Sierra and Eusebio de la Fuente, "RobHand: A Hand Exoskeleton with Real-Time EMG-Driven Embedded Control. Quantifying Hand Gesture Recognition Delays for Bilateral Rehabilitation," in *IEEE Access*, vol. 9, pp. 137809–137823, 2021, DOI: 10.1109/ACCESS.2021.3118281. Impact factor in 2021: 3.476, Q2 in "Engineering, Electrical & Electronic" (JCR-WOS).
2. **Ana Cisnal**, Paula Gordaliza, Javier Pérez-Turiel and Juan Carlos Fraile, "Interaction with a Hand Rehabilitation Exoskeleton in EMG-Driven Bilateral Therapy: Influence of Visual Biofeedback on the Users' Performance," in *Sensors*, vol. 23, 2048, 2023, DOI: 10.3390/s23042048. Impact factor in 2023: 3.4, Q2 in "Engineering, Electrical & Electronic" (JCR-WOS).
3. **Ana Cisnal**, Daniel Antolínez, Javier Pérez Turiel, Juan Carlos Fraile and Eusebio De La Fuente, "A Versatile Embedded Platform for Implementation of Biocooperative Control in Upper-Limb Neuromotor Rehabilitation Scenarios," in *IEEE Access*, vol. 11, pp. 35726–35736, 2023, DOI: 10.1109/ACCESS.2023.3265898. Impact factor in 2023: 3.4, Q2 in "Engineering, Electrical & Electronic" (JCR-WOS).

Contents

Acronyms

AAN	Assist-as-needed
ADC	Analog-to-Digital Converter
ADL	Activities of Daily Living
AFE	Analog Front End
AGD	Accurately Gesture Detection
ANN	Artificial Neural Networks
ANOVA	Analysis of Variance
ANS	Autonomic Nervous System
APB	Abductor Pollicis Brevis
ARV	Average Rectified Value
ASIC	Application Specific Integrated Circuit
BCI	Brain-Computer Interface
BLE	Bluetooth Low Energy
BP	Blood Pressure
BVP	Blood Volume Pressure
CART	Classification and Regression Trees
CDTI	Center of Development of Industrial Technology
CLA	Control Law Accelerator
CNN	Convolutional Neural Network
CVA	Cerebrovascular Accident
CVVC	Constant Voltage/Constant Current
DES	Dielectric Elastomer Sensor
DMP	Digital Motion Processor
DoF	Degrees of Freedom
DTW	Dynamic Time Warping
ECG	Electrocardiogram
ED	Extensor Digitorum
EDA	Electrodermal Activity
EE	End-Effector
EEG	Electroencephalogram
EMG	Electromyographic

EOG	Electrooculogram
(e)PWM	(enhance) Pulse Width Modulation
ERDF	European Regional Development Funds
FDS	Flexor Digitorum Superficialis
FE	Flexion/Extension
FIR	Finite Impulse Response
FN	False Negative
FP	False Positive
GBM	Gradient Boosting Machine
GSR	Galvanic Skin Response
HR	Heart Rate
HRI	Human-Robot Interaction
HRV	Heart Rate Variability
IMU	Inertial Measurement Unit
IR	Infrared thermometer
ITAP	Institute of Advanced Production Technology
JCR	Journal Citation Reports
KNN	K-Nearest Neighbors
LCSS	Longest Common Subsequence
LDA	Linear Discriminant Analysis
M3Rob	Mind-Hand-Wrist Robot
MCP	Metacarpophalangeal
MCT	Motion Completion Time
MCU	Microcontroller
ME	Motion End
MEMS	MicroElectroMechanical System
ML	Machine Learning
MO	Motion Onset
MOT	Motion Onset Time
MSPS	Mega Samples Per Second
MST	Motion Selection Time
MUAP	Motor Unit Action Potentials
MVC	Maximal Voluntary Contraction
OTP	One-Time Programmable
PCB	Printed Circuit Board
PIP	Proximal Interphalangeal
PPG	Photoplethysmogram
PPV	Positive Predictive Value
PRISMA	Reporting Items for Systematic Reviews and Meta-Analyses
PS	Pronation/Supination
QOL	Quality of Life
RESP	Respiration
RMS	Root Mean Square
RobHand	Robot for Hand Rehabilitation
ROM	Range of Motion

RR	Respiration Rate
RT	Robot-assisted Therapy
RTP	Repetitive Task Practice
RU	Radial/Ulnar
RVC	Reference Voluntary Contractions
SC	Skin Conductance
SCL	Skin Conductance Level
SCR	Skin Conductance Response
sEMG	Surface Electromyography
SKT	Skin Temperature
SMD	Surface Mounted Device
SNR	Signal-to-Noise Ratio
SOC	System-On-Chip
SVM	Support Vector Machine
TN	True Negative
TP	True Positive
TPR	True Positive Rate
VR	Virtual Reality
VRBR	Virtual Reality-Based Rehabilitation
WHO	World Health Organization
WOS	Web of Science

Chapter 1
Introduction

The incidence of cerebrovascular accidents (CVA) has been growing in the past decades as life expectancy is increasing in developed countries. Although stroke mortality has been reduced, the increase of survivors results in a rising number of adults with neurological disabilities. Restoring or improving motor skills is essential so that patients can regain independence and improve their quality of life (QOL) [1]. Although effective, traditional rehabilitation requires considerable time commitment by the rehabilitation specialist [2]. Robotic devices that allow patients to undergo rehabilitation without continuous medical assistance would make physical therapy more affordable, increasing the potential for better clinical outcomes [3].

Given the prevalence of cerebrovascular accidents, motor recovery using robotic rehabilitation systems has elicited considerable scientific interest. The active participation of the patient in the rehabilitation enhances neural plasticity and motor learning [4]. In this way, biocooperative rehabilitation robotic systems, based on multimodal information, promote patients' participation by considering their performance, motion intention and even, emotional state [5]. Despite their potential for neurorehabilitation, the first challenge to rehabilitation robotic platforms is accessibility. The presence of robotic devices in clinics and hospitals is greatly restricted since only large medical centers have the financial resources to invest in this technology [6].

In the context of neuromotor rehabilitation using biocooperative robots, a part of the budget is allocated to physiological signal acquisition systems. This research has tackled this obstacle directly by creating a cost-optimized version of biosignal recording systems, hoping to access clinical settings with more modest budgets. Its design was based on a trade-off between complexity, cost, and performance. It is necessary to provide high resolution and reliable measurements for implementing biocooperative controls in an affordable way without compromising their greatest strengths.

Biocooperative systems use biomechanical and/or physiological information to monitor patient actions, intention of movement and/or cognitive load. The emotional state of the user is not often integrated in the control loop due to the challenges

A. Cisnal de la Rica, *Development of Biocooperative Control Strategies for Neuromotor Rehabilitation Robotic Platforms*, Springer Theses,
https://doi.org/10.1007/978-3-032-02485-5_1

associated with determining it through indirect measurement of their physiological changes [7]. Motion intention can be detected by analyzing EMG signals, which are the most popular signal for implementing biocooperative controls because of its unique nature: a physiological signal that provides reliable and robust biomechanical information.

Due to the nature of EMG signals, myoelectric EMG-driven assistive robots are widely used in neuromotor rehabilitation [8, 9]. In addition, this control strategy considers the user's intention of movement, which encourages the patient to actively participate in the rehabilitation. The active participation has been proven to enhance neural plasticity and motor learning [4].

On the other hand, in the first stage of rehabilitation, it is common for the muscular electrical activity of the patient to be too weak to effectively detect their motion intention. The EMG signals can be recorded from the unimpaired limb to assist the motion of the impaired limb using the rehabilitation robot. This approach is known as bilateral myoelectric control and can replace or complement the passive therapies, which are typically performed during the first stage of rehabilitation due to the patient's inability to move the paretic limb.

Clinical studies found that passive training only reduces spasticity, while unilateral EMG-driven therapies also improved muscle coordination [10]. Furthermore, bilateral therapies have been found to be beneficial in motor recovery since the hemispheric interaction enhances the rebalancing of the abnormal brain activity caused by the stroke [11]. Even though sEMG biofeedback has been found to be beneficial in neuromotor rehabilitation [12, 13], this technique has not been used in combination with robot-assisted rehabilitation.

This thesis presents the design and development of biocooperative control strategies for upper-limb neuromotor rehabilitation, grounded in the principles of motor learning and brain plasticity. The focus on the upper limb—particularly the hand—is motivated by the high incidence of stroke-related paresis in this region [14]. Unlike traditional rehabilitation robots that rely solely on biomechanical data, the proposed strategies integrate physiological signals to enhance human–robot interaction (HRI) through a more personalized and responsive control approach. To address the limited accessibility of such systems—largely due to the high cost of commercial physiological signal acquisition devices—this work also introduces low-cost, real-time embedded platforms for biosignal recording. In doing so, the research advances both the development of human-centered control algorithms and the creation of affordable hardware solutions that support the clinical deployment of biocooperative systems.

This chapter presents a comprehensive literature review of all scientific and technical fundamentals embraced in this Doctoral Thesis. The general context is described in Sect. 1.1, which introduces neuromotor rehabilitation robotics and its clinical importance. Section 1.2 is focused on providing the basis of biocooperative controls in the context of neuromotor rehabilitation. Section 1.3 is devoted to myoelectric EMG-driven control, including different approaches for its implementation and other clinical applications based on EMG signals. Section 1.4 defines the limitations within the context of rehabilitation robotics with biocooperative control systems, while Chap. 2 establishes the central hypothesis of this work and outlines the general objective pursued in this study and enumerates the specific objectives. Chapter 3 describes the materials and methodology used in this research. Main results are presented in Chap. 4, which are further discussed in the following chapter. Finally, the contributions of this Doctoral Thesis, along with the final conclusions and future directions, are presented in the Chap. 6.

1.1 Neuromotor Rehabilitation Robotics

The World Health Organization (WHO) defined stroke as "*the neurological deficit of cerebrovascular cause that persists beyond 24 h or is interrupted by death within 24 h*" [15]. A stroke is caused by the death of brain cells due to a restricted blood flow, either by a blockage of the blood supply to a part of the brain by a clot (ischemic stroke) or by the rupture of a cerebral blood vessel (hemorrhagic stroke).

The brain is an extremely complex organ that controls many vital body functions and different abilities, such as motor and cognitive functions. Therefore, the consequences of a stroke episode depend on the type of stroke and the affected brain area, but can cause lasting brain damage, long-term neuromotor disability, or even death.

In fact, stroke is the second leading cause of death, responsible for approximately 11% of the 55.4 million worldwide deaths recorded in 2019 [16]. Additionally, it is one of the leading causes of neurological disabilities and mainly affects individuals at the peak of their productive life [17]. The rate of disability in stroke survivors is around 64%, with 30% mildly, 18% moderately, 11% severely, and 5% totally disabled (Fig. 1.1) [18]. About 70% of stroke survivors with motor disabilities require long term medical care and live with a poor quality of life (QOL) [19].

The most common and stable symptom experienced by individuals who have suffered a stroke episode is some degree of the paresis on upper extremities. Around 60% experience upper-limb dysfunction, which is especially prevalent in the hand [14]. In fact, the motor function of finger extension is usually impaired, and spasticity is often present, leading to reduced range of motion (ROM) of the hand [20]. Muscle weakness is also exhibited to varying degrees in most of stroke survivors [21, 22].

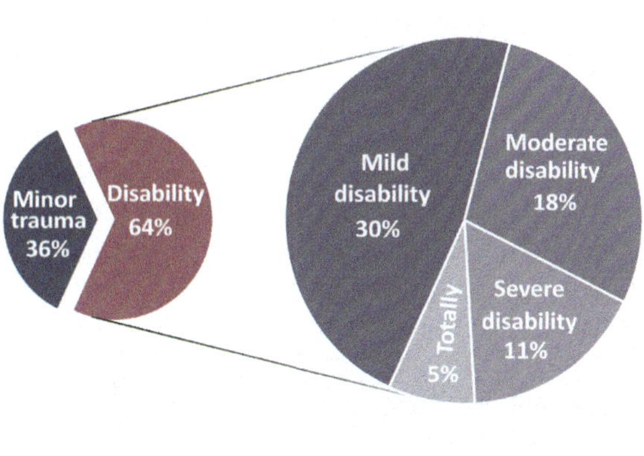

Fig. 1.1 Rate of disability in stroke survivors, broken down by degree of disability

Loss of hand function is a major source of impairment in neuromotor disorders, as it is essential for manipulating the environment and thus worsens the abilities of stroke survivors in performing activities of daily living (ADL). Rehabilitation based on physical therapy is the primary mechanism for improving motor function and achievement of independent participation in daily life [23]. Early and intensive physical rehabilitation is of great importance to maximize functional motor recovery [24, 25].

Repetitive task practice (RTP) rehabilitation consists of breaking ADLs down into individual movements and continuously and intensively practicing them. It has been proven that it promotes hand functional motor recovery, especially leading to improved range of motion and strength [26]. However, RTP rehabilitation costs are very high due to the heavy workload of the rehabilitation specialists [2].

Additionally, the aging population is linked to an increase in stroke survivors with neuromotor disabilities and a considerable reduction in the patient-physician ratio. Interest in robotic rehabilitation systems has experienced considerable growth in the last decade not only because of their high costs, but also to increase rehabilitation capacity worldwide and offer adequate rehabilitation to all patients [27]. A system that allows patients to intensively rehabilitate without the continuous assistance of the therapist, would make physical therapy more affordable and accessible.

Clinical studies have shown improvement in hand motor function when performing robot-assisted therapy (RT) [28–31]. Although the rehabilitation benefits of RT in upper-limb motor recovery of people after stroke is not significantly better than those obtained with conventional rehabilitation, RT offers major advantages in terms of lower manpower cost [11, 32]. Furthermore, it improves rehabilitation convenience facilitating independent therapy with the potential outcome of increasing intensity and patient motivation [33, 34].

Rehabilitation robotic devices can be mainly divided into two categories based on their structural design: exoskeletons and end-effectors (EE). Although clinical results are limited, it has been found that exoskeleton robotic devices provide better rehabilitation outcomes than EE systems in improving hand motor impairments [35].

Among these, upper-limb rehabilitation exoskeletons are generally more technologically advanced compared to hand exoskeletons, largely due to the anatomical complexity of the hand. Specifically, the human hand compromises 21 DOFs, whereas the arm (from the wrist to the shoulder) has only 7 DoFs. Designing a hand exoskeleton involves addressing multiple challenges, such as minimizing size and weight while ensuring dexterous manipulation capabilities. Additional considerations include determining the appropriate number of degrees of freedom, which joints to actuate directly, and which grasp patterns to prioritize or renounce [36].

Several robotic systems for hand rehabilitation are already commercially available. These includes devices such as Saeboflex (Saebo, Inc., NC, USA), the Motus hand (Motus Nova, GA, USA), Hand of Hope (Rehab-Robotics, China), Amadeo (TyroMotion, Austria), GLOREHA (Idrogenet, Italy) and Waveflex CMP (Remington Medical, Canada) and HandyRehab (Fourier Intelligence, Singapore).

Beyond the cost-effectiveness and intensive training offered by robotic therapy (RT), these systems support different training paradigms, which are used depending

Table 1.1 Training modalities for rehabilitation robotics, categorized based on the interaction between the rehabilitation robot and the patient

Training modalities	Patient		Robot
	Intention	Force	Force
Passive	0	0	+++
Assistive	+	+	+
Active-assistive	++	+/0	0/+
Active	++	+	0
Resistive	++	++	–

on the stroke stage and patient's level of motor function [37]. A training paradigm defines the interaction between the rehabilitation robot and the patient, taking into account the patient's movement intentions and the forces Applied [38]. These training paradigms are known as passive, assistive, active-assistive, active, and resistive (Table 1.1).

After suffering a stroke, the person may completely lose mobility of the paretic limb due to muscle spasticity. In the first stage of rehabilitation, passive and active-assistive exercises are optimal since the robot assists the movement of the paretic limb without requiring any motion ability of the patient. Passive therapies have been found to be temporarily effective for reducing hypertonia [39] and for maintaining the ROM of the hand in the early stage of the treatment, but they do not significantly improve motor function [40]. Active-assistive exercises were found to be more effective on motor recovery than passive control since active engagement of the patient is required [4].

In the second stage of the stroke recovery, active and assistive exercises are normally used when the patient has motion capabilities. The robot provides assisting forces to support the patient in completing the desired movement. In the later period of stroke, resistive exercises are used to increase muscle force because the user must complete the motion against a resistive force exerted by the robot [41].

Evidence has shown that the active participation of the patient induces neural plasticity in motor learning. Hence, robotic systems should monitor the patient's intention and promote patient participation to optimize the therapy outcomes. This can be achieved by providing adaptive exercises or assistance based on patient motion intent. Assist-as-needed (AAN) algorithms are based on providing the minimum necessary motion assistance, so the patient can complete a movement, thus requiring significant motion from the patient. In the case of severely impaired patients who are unable to move the limb, active-assistive exercises based on intention detection recognition encourage them to make an effort to move their paretic limb [4].

User motion and motor intention are normally detected by measuring position or contact forces [42]. Additionally, motor intention can also be detected from brain cortical activity by analyzing EEG signals and from the muscles by measuring their electrical activity or EMG signals. The electrical activity can be detected from the paretic hand or the contralateral healthy hand [38]. In fact, using the electrical activity from the unimpaired side to identify user intention is recommended in the case of patients with severe hand paresis or insufficient muscle tone [1]. This approach is known as bilateral-assisted therapy.

The robotic device provides motion assistance by referring to detected or intended movement of an unimpaired counterpart. Some studies suggest that bilateral therapies are potentially more effective than unilateral exercise in upper-limb rehabilitation. Motor recovery of bilateral therapies is based on hemispheric interaction. It increases the excitability of the motor complex and supplementary motor area, resulting in rebalancing the abnormal interhemispheric activities caused by stroke. Additionally, bilateral exercises improve inter-limb coordination [11].

One of the primary challenges of neuromotor rehabilitation is maintaining motivation over long periods of time required for significant and lasting functional improvement. RT allows for the integration of virtual reality (VR) environments into the rehabilitation process. Within virtual reality-based rehabilitation (VRBR), three key concepts play a vital role in motor learning: repetition, feedback, and motivation. By offering repetitive practice, augmented feedback, and motivation to endure practice, VR serves as a powerful tool to enhance motor learning and induce cortical and subcortical changes associated with skilled tasks [43].

VRBR offers patients the opportunity to benefit from the robotic rehabilitation experience via serious games. Stroke patients have shown improved emotional responses to VRBR [44]. It provides motivation for post-stroke patients [45], potentially increasing rehabilitation intensity and consequently, promoting motor recovery [46]. Ideally, stroke rehabilitation games should be based on ADLs, while ensuring active patient participation [47]. Additionally, incorporating multisensory feedback enhances connectivity between sensory and motor cortices, further promoting motor learning [48].

The feedback from serious games, such as the score, provides understandable indicators of the patient's performance and stimulates the learning process. Reinforcement learning is one of the most powerful ways of learning new skills. This stimulation of the learning process is based on dopamine release in the key areas of the brain. It leads to improved consolidation of long-term motor memory and strengthens motivation [49]. Hence, providing a rate of recovery feedback encourages the patient to perform better next session. In general, the use of VRBR has been found beneficial for stroke patients [50].

1.2 Biocooperative Control in Rehabilitation Robotics

The concept of biocooperative control emerged in the field of medical robotics, specifically in rehabilitation robotics during the early years of the twenty-first century. One of the first formal appearances of the term "biocooperative control" was in a special section of the IEEE Transactions on Neural Systems and Rehabilitation Engineering in 2010, where the basis of biocooperative systems was outlined and formally defined as those that introduced physiological or psychological information into the control loop [5].

The first rehabilitation robots were based on an open-loop control with fixed position and/or velocity references to execute a predefined trajectory to assist in the patient's movement. This technique did not involve either biomechanical or psycho-physiological information. Consequently, these systems were unable to respond to a patient's voluntary effort or spontaneous intentions, and efficient human–machine interaction was practically impossible. In the first decade of this century, rehabilitation robots appeared that employed biomechanical data, such as position, velocity, acceleration, and force to establish robust and adaptive controls. The main goal was to achieve safe, ergonomically acceptable, and user-cooperative systems by controlling the biomechanical interaction between the robot and the patient. This new approach no longer considers the patient as a disturbance applied directly to the robotic system and provides a bidirectional interaction between the robot and the patient enhancing the rehabilitation experience [51].

The biocooperative approach in rehabilitation robotics goes one step further and integrates the patient into the feedback loop by not only considering biomechanical information but also including psychological and/or physiological measurements. Hence, the appearance of biocooperative controls has led to the emergence of a new generation of rehabilitation robotic platforms, which record and control the patient's physiological signals. They are usually based on multimodal interfaces; information coming from different sources allows for continuously monitoring the patient global status including their actions, intention of movement, emotional state, and even environmental factors. Figure 1.2 shows a generic diagram of biocooperative robotic systems, specifying the integration of the human into the loop in a biomechanical, physiological, and psychological sense.

Emotions are associated with physiological changes produced in response to the autonomic nervous system (ANS). Facial expressions, breathing patterns, heart rate or muscle tension changes are influenced by emotions. Physiological measurements such as cardiovascular parameters, temperature, electrodermal activity, blood

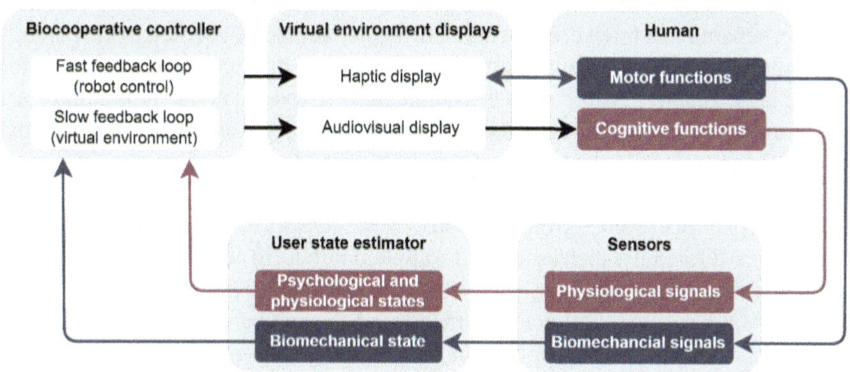

Fig. 1.2 Generic diagram of biocooperative robotic systems, specifying integration of the human into the loop in a biomechanical, physiological, and psychological sense [5]

oxygenation, photoplethysmography and electromyography have been used to indirectly measure ANS-related responses to external stimuli and hence, determine the emotional state of the person. Although other biomarkers such as cortisol levels or neural changes revealed by neuroimaging can also reveal useful information regarding the emotional state [52], they are potentially invasive and inconvenient for rehabilitation robotics.

In this field, a diverse number of non-invasive biological signals such as EMG, EEG, EOG, RESP, GSR, SKT and BP are commonly employed to extract valuable physiological information from the user. Signal processing techniques are applied to extract specific features from these signals that are known to exhibit correlations with the user's physiological state. For instance, parameters such as heart rate (HR) and heart rate variability (HRV) derived from ECG signals, and the skin conductance level (SCL) and the frequency of the skin conductance response (SCR) derived from GSR signals, serve as indicators of arousal, physical effort, and mental workload. Consequently, these biosignals allow determining the cognitive load and physiological response of the user to the rehabilitation session. Additionally, physiological signals can also be used to detect the user intention through non-invasive brain-computer interfaces (BCIs) based on EEG and non-cortical interfaces based on EOG or EMG [53]. These interfaces are particularly relevant in the case of patients with severe impairments, as they are able to generate muscle activation or brain activity instead of force for movement [54].

Table 1.2 shows a summary of biocooperative controls in the context of neurorehabilitation proposed in the literature. In healthy individuals, motor intention is conducted from the brain to the muscle to directly move the limb. However, this natural pathway is damaged in stroke patients and motion intention is detected based on brain or muscle activity or even eye movement in order for the robot to actuate the motion of the impaired limb. Hence, motion intention recognition (MR in Table 1.2) based on EEG, EMG or EOG is a common approach in biocooperative robots.

Control based on the AAN paradigm is widely used in rehabilitation robotics since personalized difficulty has been proven to lead to superior learning outcomes when compared to a fixed difficulty. The difficulty should be adapted to the patient's capabilities without surpassing them as this could have detrimental effects on performance. The adaptive nature of task complexity increases activity in premotor and sensorimotor areas and has beneficial effects on motor recovery [48]. In this context, an AAN control architecture based solely on myoelectric information from the muscles (EMG-ANN in Table 1.2) has been proposed. Biomechanical information of the user typically extracted from position or force sensors is replaced by information based on EMG signals, such as the user's applied torque, to adapt the assistance level of the rehabilitation robot.

More complex AAN architectures in terms of the number of physiological signal modalities have also been proposed in the literature (BIO-AAN in Table 1.2). The level of assistance or task difficulty is adapted according to the user's performance, muscle fatigue, or emotional state. While user performance is normally determined by biomechanical variables, muscle fatigue is determined by analyzing EMG signals.

Table 1.2 Literature review on biocooperative systems, indicating the control type and the employed physiological signals

Ref	Type	EOG	EEG	ECG	EMG	IMU/ACC	GSR	SKT	RESP	BVP	Other
[55]	MR	x	x		x						
[56]	MR				x	x					
[57]	MR				x						x[1]
[58]	MR				x	x					
[59]	EMG-AAN				x						
[60]	EMG-AAN				x						
[61]	EMG-AAN				x						
[62]	BIO-AAN				x	x					
[63]	BIO-AAN			x			x	x	x		
[64]	BIO-AAN						x	x	x		
[65]	BIO-AAN										
[66]	AAN						x	x	x		
[67]	ES		x				x	x		x	
[68]	ES	x		x	x		x			x	
[69]	ES		x				x				
[70]	ES	x	x		x		x	x	x		
[71]	ES			x	x		x		x		
[72]	ES			x	x		x				
[73]	ES			x	x		x	x		x	x[2]
[74]	ES				x		x		x	x	
[75]	ES	x					x			x	

[1]DES (Dielectric Elastomer Sensor)
[2]Bioimpedance, heart sound and PPG (photoplethysmogram)

The emotional response of the user, including the level of arousal, valence, or stress, is also included in the control loop.

Other works have mainly focused on the emotional estimation (ES in Table 1.2) using multimodal information. The development of emotion recognition algorithms, also known as affective computing, are still a major challenge. Efficient and robust affective computing would significantly increase the quality of human–machine interaction, not only in the context of neuromotor rehabilitation, but also in other fields ranging from preventive medicine to the multimedia industry [7]. Physiological parameters such as HR, SKT, electrodermal activity (EDA), and respiration rate (RR) are known to be associated with autonomic responses related to emotions and are therefore often employed as inputs for emotion classifiers.

Biocooperative controls allow the implementation of more advanced and personalized AAN control paradigms than traditional ones that only rely on user performance based on biomechanical information. These controls also employ other physiological signals to consider additional variables, such as cognitive load, physiological state, or even muscle fatigue. This is of great importance since negative emotions such as anxiety, frustration, or stress can have a significant impact on motor learning. Despite the proven benefits of personalized adaptive controls in facilitating motor recovery, the advancement of biocooperative controls has been hindered by a slow progress. This may be due to the complexity of developing robust and efficient algorithms that analyze the human body's physiological response to ANS-triggered stimuli and determine emotions, combined with the high cost and bulky nature of physiological signal recording systems.

Most of the reviewed studies relied on expensive, bulky, and computationally limited commercial products such as the MP150 system (BIOPAC, CA, USA), the Neuroscan NuAmps Express system (Compumedics Ltd., Australia), the ActiveTwo system (Biosemi, Netherlands), and the Pro-Comp/FlexComp Infiniti system (Thought Technology Ltd., Canada). Only a few studies utilized specific wearable technologies from Biometrics Ltd (UK), g.tec medical engineering GmbH (Austria), or Delsys Incorporated (MA, USA).

1.3 Myoelectric EMG-Driven Control

Electromyography (EMG), sometimes referred to as myoelectric activity, is a technique that focuses on the development, recording, and analysis of the information present in the electric potential generated by motor units. Motor units are the smallest functional units that describe the neural control of the muscular contraction process when the muscle is activated [76].

EMG is classified into intramuscular and surface EMG (sEMG). While intramuscular EMG uses tiny needles or fine wires that are inserted into the muscle through the skin, sEMG uses surface electrodes. sEMG signals represent the electrically superimposed motor unit action potentials (MUAPs) of all active motor units detectable under the electrode site (Fig. 1.3). The main advantages of sEMG are that it is non-invasive, easy to record, and allows for the estimation of overall muscle activity. The disadvantage of sEMG compared to intramuscular EMG is that it is only suitable for recording superficial muscles and may experience possible contamination from nearby muscles, especially when recording small muscles. In contrast, intramuscular EMG is selective and enables recording of deep muscle activity [77].

Although sEMG signals are often considered more complex to analyze than intramuscular EMG, they tend to be used in rehabilitation medicine applications due to their non-invasive nature. Additionally, their analysis provides a global measurement of the level of muscle activity, which may be more appropriate in movement analysis.

In fact, sEMG has been regarded as the most suitable physiological signal for implementing biocooperative controls due to its good accuracy and robustness in predicting the intention of human motion [8]. Analyzing sEMG signals from the target muscles is widely used as an alternative strategy for integrating force sensors into the rehabilitation robot for motion control. Force sensors have intrinsic problems,

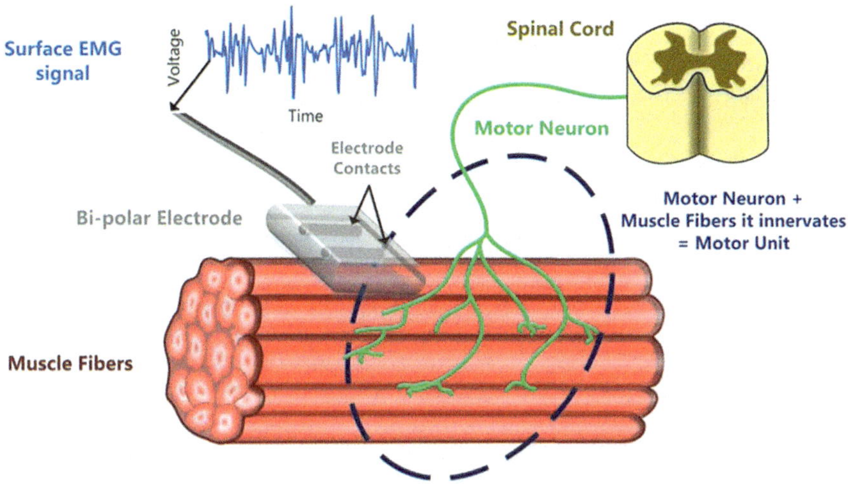

Fig. 1.3 EMG signal recording using a surface electrode, including a motor unit and muscle fibers [77]

such as their placement, which can obstruct the sense of feeling, or the inability to distinguish between user-applied and external forces. Robotic control systems based on EMG signals are the most popular approach in rehabilitation robotics [9].

EMG-based motion intention recognition can be broadly categorized as classification and regression problems. While classification-based myoelectric control detects the type of movement, regression-based control can output continuous variables such as joint angle and joint torque. Therefore, the first strategy enables discrete motion control and is limited to a specific number of motions, whereas the latter allows for continuous motion control, resembling the continuous movement of the human body [78]. Despite this fact, classification-based myoelectric controls are still the most widely implemented due to their reliable results. In both approaches, the kinematic parameters estimated from the analysis of the EMG signals are fed into the robot controller, causing it to behave according to the human intention (see Fig. 1.4).

Regarding discrete motion control, it can be classified into non-pattern recognition and pattern recognition methods. Non-pattern recognition methods, also known as threshold methods, generate the input commands for the robot's assistance controller by comparing the amplitude or statistical features of EMG signals with some thresholds. Pattern recognition methods detect the type of movement using classification machine learning models, such as support vector machines (SVM), artificial neural networks (ANN), linear discriminant analysis (LDA), among others. The classification results are used as an input to the robot controller. Non-pattern recognition-based myoelectric control methods have the lowest computational cost [79].

In contrast, the continuous motion control outputs variables such as joint angle or joint torque from the analysis of EMG signals. The use of biomechanical and machine learning (ML) models is the main method for mapping the EMG signals to

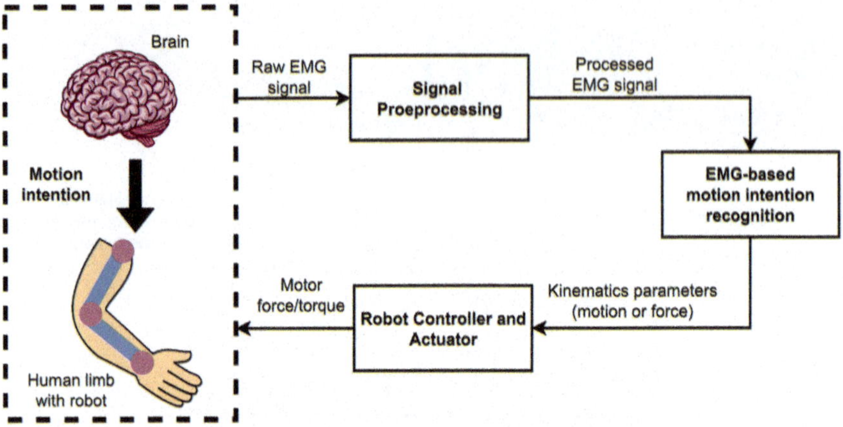

Fig. 1.4 Conceptual block diagram for myoelectric control system

the continuous control input. The most widely used biomechanical model is Hill's Muscle Model, which represents the human skeletal muscle with a 3-element system: a contractile element and two non-linear spring elements, one in series and another in parallel (Fig. 1.5) [80]. On the other hand, time-domain or statistical features extracted from the EMG signals are often used as inputs to regression models, such as linear regression or ANN, to estimate joint angle or applied torque [79]. However, ML-based systems have not been implemented in real applications mainly because of reliability issues [9].

The potential applications of myoelectric control extend beyond motor recovery in rehabilitation; they also include assistance in ADL and human augmentation to enhance physical capacities of the healthy population [79]. In the case of rehabilitation robotics, the choice of the type of EMG control must not only consider the DoFs of the assistive device, but also user-acceptance in terms of the number of electrodes and the system response. For instance, a high number of electrodes allows the implementation of an accurate continuous EMG control at the cost of poorer usability and a higher computational cost, which may be reflected in the system response. In their review [79], Fu et al. found that 42% of the articles focused on controlling only one DoF, while 15%, 3%, 13% and 15% of the studied articles integrated two, three, four and more than four DoFs, respectively.

Clinical studies have been undertaken to evaluate the effectiveness of EMG-driven therapies in stroke patients. They found that while passive training mainly reduced spasticity, EMG-driven therapies also improved muscle coordination [10]. However, some stroke patients exhibit sEMG signals that are excessively weak to effectively detect their motion intention. In such cases, a popular strategy is the implementation of bilateral myoelectric control, which consists of estimating the motion intention using EMG signals recorded from an unimpaired limb to rehabilitate the impaired limb with the assistance of a robotic device.

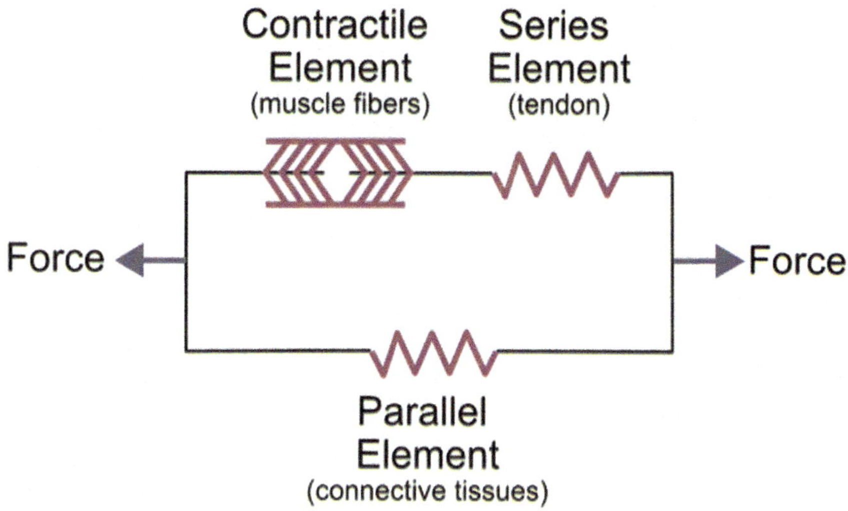

Fig. 1.5 Schematic diagram of the 3-element Hill model of the human skeletal muscle: a contractile element, series element, and parallel element modelling acting and muscle fibers, tendons, and connective tissues, respectively

Moreover, sEMG can serve as an indirect measure to identify changes in muscle force during isometric muscle contractions, making it a potential tool for assessing muscle fatigue objectively. It can also be used as real-time feedback to make patients more aware of their muscle activity and support re-learning of movement patterns. The amplitude of the sEMG signals can be fed back to the user to provide an objective measure of the level of muscle activation [77]. Biofeedback has been generally delivered using visual displays or acoustic signals.

The biofeedback technique was introduced more than forty years ago in rehabilitation settings [13]. sEMG biofeedback has been found to be beneficial in the treatment of many musculoskeletal conditions, including stroke rehabilitation, pelvic floor muscle dysfunction, and even in alleviating pain caused by muscular tension [12]. EMG biofeedback allows users to learn to self-regulate their muscle activity and facilitate the process of strengthening weak or paretic muscles and reducing tone in spastic muscles. However, clinical studies that have assessed the effectiveness of EMG biofeedback in musculoskeletal and neuromotor rehabilitation did not include the use of this technique in combination with RT.

1.4 Current Limitations

The concept of utilizing robotic devices for neurorehabilitation was initially introduced by Hogan at MIT [81]. Despite the gradual introduction of traditional rehabilitation robots into clinical settings, the current level of development of biocooperative rehabilitation robots is insufficient to transition them from laboratory settings to practical real-world applications. While biocooperative controls promoting active participation that has been proven to enhance neural plasticity and motor [4], their high cost does not provide sufficient incentives for extensive investigation, hindering their current application and commercialization. Therefore, in order to advance the research and facilitate the utilization of biocooperative systems in clinical environments, it is necessary to address, at minimum, the following primary limitations:

I. Hardware

Recent research emphasizes the importance of compact, affordable sensing devices with advanced computational capabilities for detecting human physical activity and emotions through multimodal fusion strategies [82]. Despite the decreasing cost of computing and the emergence of real-time control support through embedded solutions [83], biocooperative systems still rely on expensive, computationally limited, and occasionally bulky commercially available acquisition systems [84], which impacts on accessibility and user acceptance.

II. Reliability

Biocooperative controls currently lack the required reliability and robustness to be effectively implemented in real-world scenarios. This includes the specific control method based on EMG signals, which, despite being widely utilized in the field of neuromotor rehabilitation robotics due to their accurate and robust prediction of motion intentions [8], have not been successfully applied in clinical practice due to reliability concerns [9].

References

1. Tran P, Jeong S, Herrin KR, Desai JP (2021) Review: hand exoskeleton systems, clinical rehabilitation practices, and future prospects. IEEE Trans Med Robot Bionics 3:606–622. https://doi.org/10.1109/tmrb.2021.3100625
2. Frick EM, Alberts JL (2007) Combined use of repetitive task practice and an assistive robotic device in a patient with subacute stroke. Phys Ther 86:1378–1386. https://doi.org/10.2522/ptj.20050149
3. Polygerinos P, Wang Z, Galloway KC, Wood RJ, Walsh CJ (2015) Soft robotic glove for combined assistance and at-home rehabilitation. Rob Auton Syst 73:135–143. https://doi.org/10.1016/j.robot.2014.08.014
4. Blank AA, French JA, Pehlivan AU, O'Malley MK (2014) Current Trends in robot-assisted upper-limb stroke rehabilitation: promoting patient engagement in therapy. Curr Phys Med Rehabil Rep 2:184–195. https://doi.org/10.1007/s40141-014-0056-z

5. Riener R, Munih M (2010) Guest editorial special section on rehabilitation via bio-cooperative control. IEEE Trans Neural Syst Rehabil Eng 18:337–338. https://doi.org/10.1109/TNSRE.2010.2060390
6. Almekkawy M, Zderic V, Chen J, Ellis MD, Haemmerich D, Holmes DR, Linte CA, Panescu D, Pearce J, Prakash P (2020) Therapeutic systems and technologies: state-of-the-art applications, opportunities, and challenges. IEEE Rev Biomed Eng 13:325–339. https://doi.org/10.1109/RBME.2019.2908940
7. Katsigiannis S, Ramzan N (2018) DREAMER: a database for emotion recognition through EEG and ECG Signals from wireless low-cost off-the-shelf devices. IEEE J Biomed Health Inform 22:98–107. https://doi.org/10.1109/JBHI.2017.2688239
8. Li K, Zhang J, Wang L, Zhang M, Li J, Bao S (2020) A review of the key technologies for sEMG-based human-robot interaction systems. Biomed Signal Process Control 62:102074. https://doi.org/10.1016/j.bspc.2020.102074
9. Meattini R, Benatti S, Scarcia U, De Gregorio D, Benini L, Melchiorri C (2018) An sEMG-based human-robot interface for robotic hands using machine learning and synergies. IEEE Trans Compon Packag Manuf Technol 8:1149–1158. https://doi.org/10.1109/TCPMT.2018.2799987
10. Hu XL, Tong KY, Song R, Zheng XJ, Leung WWF (2009) A comparison between electromyography-driven robot and passive motion device on wrist rehabilitation for chronic stroke. Neurorehabil Neural Repair 23:837–846. https://doi.org/10.1177/1545968309338191
11. Wu J, Cheng H, Zhang J, Yang S, Cai S (2021) Robot-assisted therapy for upper extremity motor impairment after stroke: a systematic review and meta-analysis. Phys Ther 101:1–13. https://doi.org/10.1093/ptj/pzab010
12. Giggins OM, Persson UM, Caulfield B (2013) Biofeedback in rehabilitation. J NeuroEngineering Rehabil 10:1–11. https://doi.org/10.1186/1743-0003-10-60
13. Tate JJ, Milner CE (2010) Real-time kinematic, temporospatial, and kinetic biofeedback during gait retraining in patients: a systematic review. Phys Ther Rehabil J 90:1123–1134. https://doi.org/10.2522/ptj.20080281
14. Fischer HC, Stubblefield K, Kline T, Luo X, Kenyon RV, Kamper DG (2007) Hand rehabilitation following stroke: a pilot study of assisted finger extension training in a virtual environment. Top Stroke Rehabil 14:1–12. https://doi.org/10.1310/tsr1401-1
15. World Health Organization (1978) Cerebrovascular disorders: a clinical and research classification. World Health Organization, Geneva
16. World Health Organization (2020) The top 10 causes of death [WWW Document]. WHO. URL https://www.who.int/news-room/fact-sheets/detail/the-top-10-causes-of-death. Accessed 2 Aug 2023
17. GBD 2016 Stroke Collaborators (2019) Global, regional, and national burden of stroke, 1990–2016: a systematic analysis for the global burden of disease study 2016. Lancet Neurol 18:439–458. https://doi.org/10.1016/S1474-4422(19)30034-1
18. Lv Y, Sun Q, Li J, Zhang W, He Y, Zhou Y (2021) Disability status and its influencing factors among stroke patients in northeast China: A 3-year follow-up study. Neuropsychiatr Dis Treat 17:2567. https://doi.org/10.2147/NDT.S320785
19. Parker ME, Snyder-Shall M (2013) Recovery of upper extremity function in stroke patients. Neurol Rep 19:46–47. https://doi.org/10.1097/01253086-199519010-00028
20. Kamper DG, Harvey RL, Suresh S, Rymer WZ (2003) Relative contributions of neural mechanisms versus muscle mechanics in promoting finger extension deficits following stroke. Muscle Nerve 28:309–318. https://doi.org/10.1002/mus.10443
21. Ada L, Canning CG, Low SL (2003) Stroke patients have selective muscle weakness in shortened range. Brain 126:724–731. https://doi.org/10.1093/brain/awg066
22. Colebatch JG, Gandevia SC (1989) The distribution of muscular weakness in upper motor neuron lesions affecting the arm. Brain 112:749–763. https://doi.org/10.1093/brain/112.3.749
23. Winstein CJ, Stein J, Arena R, Bates B, Cherney LR, Cramer SC, Deruyter F, Eng JJ, Fisher B, Harvey RL, Lang CE, MacKay-Lyons M, Ottenbacher KJ, Pugh S, Reeves MJ, Richards LG, Stiers W, Zorowitz RD (2016) Guidelines for adult stroke rehabilitation and recovery:

a guideline for healthcare professionals from the American heart association/american stroke association. Lippincott Williams and Wilkins, Stroke. https://doi.org/10.1161/STR.0000000000000098
24. Barria P, Riquelme M, Reppich H, Cisnal A, Fraile JC, Pérez-Turiel J, Sierra D, Aguilar R, Andrade A, Nuñez-Espinosa C (2023) Hand rehabilitation based on the RobHand exoskeleton in stroke patients: a case series study. Front Robot AI 10:1–15. https://doi.org/10.3389/frobt.2023.1146018
25. Herpich F, Rincon F (2020) Management of acute ischemic stroke. Crit Care Med 48:1654–1663. https://doi.org/10.1097/CCM.0000000000004597
26. Sterr A, Freivogel S (2003) Motor-improvement following intensive training in low-functioning chronic hemiparesis. Neurology 61:842–844. https://doi.org/10.1212/wnl.61.6.842
27. Cisnal A, Alonso-Linaje G, Fraile JC, Pérez-Turiel J, Álvarez P, Martinez S (2024) Tackling post-COVID-19 rehabilitation challenges: a pilot clinical trial investigating the role of robotic-assisted hand rehabilitation. J Clin Med 13. https://doi.org/10.3390/jcm13061543
28. Carmeli E, Peleg S, Bartur G, Elbo E, Vatine J-J (2011) HandTutorTM enhanced hand rehabilitation after stroke—a pilot study. Physiother Res Int 16:191–200. https://doi.org/10.1002/pri.485
29. Kutner NG, Zhang R, Butler AJ, Wolf SL, Alberts JL (2010) Quality-of-life change associated with robotic-assisted therapy to improve hand motor function in patients with subacute stroke: a randomized clinical trial. Phys Ther 90:493–504. https://doi.org/10.2522/ptj.20090160
30. Ueki S, Nishimoto Y, Abe M, Kawasaki H, Ito S, Ishigure Y, Mizumoto J, Ojika T (2008) Development of virtual reality exercise of hand motion assist robot for rehabilitation therapy by patient self-motion control. In: 2008 30th Annual international conference of the IEEE engineering in medicine and biology society. IEEE, Vancouver, British Columbia, Canada, pp 4282–4285. https://doi.org/10.1109/IEMBS.2008.4650156
31. Wolf SL, Winstein CJ, Miller JP, Taub E, Uswatte G, Morris D, Giuliani C, Light KE, Nichols-Larsen D, EXCITE Investigators (2006) Effect of constraint-induced movement therapy on upper extremity function 3 to 9 months after stroke: the EXCITE randomized clinical trial. JAMA 296:2095–2104. https://doi.org/10.1001/jama.296.17.2095
32. Chien WT, Chong YY, Tse MK, Chien CW, Cheng HY (2020) Robot-assisted therapy for upper-limb rehabilitation in subacute stroke patients: a systematic review and meta-analysis. Brain Behav 10:1–16. https://doi.org/10.1002/brb3.1742
33. Kwakkel G, Kollen BJ, Krebs HI (2008) Effects of robot-assisted therapy on upper limb recovery after stroke: a systematic review. Neurorehabil Neural Repair 22:111–121. https://doi.org/10.1177/1545968307305457
34. Rietman JS, Prange G, Kottink A, Ribbers G, Buurke J (2014) The effect of an arm supporting training device in sub-acute stroke patients: randomized clinical trial. Arch Phys Med Rehabil 95:e8. https://doi.org/10.1016/j.apmr.2014.07.382
35. Moggio L, de Sire A, Marotta N, Demeco A, Ammendolia A (2022) Exoskeleton versus end-effector robot-assisted therapy for finger-hand motor recovery in stroke survivors: systematic review and meta-analysis. Top Stroke Rehabil 29:539–550. https://doi.org/10.1080/10749357.2021.1967657
36. Lum PS, Godfrey SB, Brokaw EB, Holley RJ, Nichols D (2012) Robotic approaches for rehabilitation of hand function after stroke. Am J Phys Med Rehabil 91:242–251. https://doi.org/10.1097/PHM.0b013e31826bcedb
37. Kahn LE, Lum PS, Rymer WZ, Reinkensmeyer DJ (2006) Robot-assisted movement training for the stroke-impaired arm: does it matter what the robot does? J Rehabil Res Dev 43:619–630
38. Yue Z, Zhang X, Wang J (2017) Hand rehabilitation robotics on poststroke motor recovery. Behav Neurol (2017):1–20. https://doi.org/10.1155/2017/3908135
39. Schmit BD, Dewald JPA, Rymer WZ (2000) Stretch reflex adaptation in elbow flexors during repeated passive movements in unilateral brain-injured patients. Arch Phys Med Rehabil 81:269–278. https://doi.org/10.1016/S0003-9993(00)90070-4
40. Volpe BT, Krebs HI, Hogan N, Edelstein L, Diels C, Aisen M (2000) A novel approach to stroke rehabilitation: robot-aided sensorimotor stimulation. Neurology 54:1938–1944. https://doi.org/10.1212/WNL.54.10.1938

41. Fasoli SE, Krebs HI, Stein J, Frontera WR, Hogan N (2003) Effects of robotic therapy on motor impairment and recovery in chronic stroke. Arch Phys Med Rehabil 84:477–482. https://doi.org/10.1053/apmr.2003.50110
42. Cisnal A, Pérez-Turiel J, Fraile JC (2025) Enhancing hand robotic rehabilitation with bilateral therapy using leap motion controller. In: Pons JL, Tornero J, Akay M (eds) Converging clinical and engineering research on neurorehabilitation V. Springer Nature Switzerland, Cham, pp 379–383. https://doi.org/10.1007/978-3-031-77588-8_76
43. Holden MK (2005) Virtual environments for motor rehabilitation: review. Cyberpsychol Behav 8:187–211. https://doi.org/10.1089/cpb.2005.8.187
44. Cisnal A, Moreno-SanJuan V, Fraile JC, Turiel JP, de-la-Fuente E, Sánchez-Brizuela G (2022) Assessment of the patient's emotional response with the RobHand rehabilitation platform: a case series study. J Clin Med 11:4442. https://doi.org/10.3390/jcm11154442
45. Reinkensmeyer DJ, Housman SJ (2007) If I can't do it once, why do it a hundred times? Connecting volition to movement success in a virtual environment motivates people to exercise the arm after stroke. In: 2007 Virtual rehabilitation. IEEE, Venice, Italy, pp 44–48. https://doi.org/10.1109/ICVR.2007.4362128
46. Corbetta D, Imeri F, Gatti R (2015) Rehabilitation that incorporates virtual reality is more effective than standard rehabilitation for improving walking speed, balance and mobility after stroke: a systematic review. J Physiother 61:117–124. https://doi.org/10.1016/J.JPHYS.2015.05.017
47. Sveistrup H (2004) Motor rehabilitation using virtual reality. J Neuroeng Rehabil 1:1–8. https://doi.org/10.1186/1743-0003-1-10
48. Maier M, Ballester BR, Verschure PFMJ (2019) Principles of neurorehabilitation after stroke based on motor learning and brain plasticity mechanisms. Front Syst Neurosci 13:1–18. https://doi.org/10.3389/fnsys.2019.00074
49. Bo Nielsen J, Willerslev-Olsen M, Christiansen L, Lundbye-Jensen J, Lorentzen J (2015) Science-based neurorehabilitation: Recommendations for neurorehabilitation from basic science. J Mot Behav 47:7–17. https://doi.org/10.1080/00222895.2014.931273
50. Subramanian SK, Lourenço CB, Chilingaryan G, Sveistrup H, Levin MF (2013) Arm motor recovery using a virtual reality intervention in chronic stroke: randomized control trial. Neurorehabil Neural Repair 27:13–23. https://doi.org/10.1177/1545968312449695
51. Koenig A, Omlin X, Novak D, Riener R (2011) A review on bio-cooperative control in gait rehabilitation. In: IEEE international conference on rehabilitation robotics. IEEE, Zurich, pp 1–6. https://doi.org/10.1109/ICORR.2011.5975454
52. Jerath R, Beveridge C (2020) Respiratory rhythm, autonomic modulation, and the spectrum of emotions: the future of emotion recognition and modulation. Front Psychol 11:1–7. https://doi.org/10.3389/fpsyg.2020.01980
53. Antolínez D, Cisnal A, Martínez-Cagigal V, Santamaría-Vázquez E, Benavides D, Granja J, Fraile JC, Hornero R (2025) Development and validation of an affordable wearable multichannel EEG system with active electrodes and innovative flexible headbands. IEEE Access 13:83983–83995. https://doi.org/10.1109/ACCESS.2025.3568302
54. Simonetti D, Zollo L, Papaleo E, Carpino G, Guglielmelli E (2016) Multimodal adaptive interfaces for 3D robot-mediated upper limb neuro-rehabilitation: an overview of bio-cooperative systems. Rob Auton Syst 85:62–72. https://doi.org/10.1016/j.robot.2016.08.012
55. Zhang J, Wang B, Zhang C, Xiao Y, Wang MY (2019) An EEG/EMG/EOG-based multimodal human-machine interface to real-time control of a soft robot hand. Front Neurorobot 13:1–13. https://doi.org/10.3389/fnbot.2019.00007
56. Krasoulis A, Vijayakumar S, Nazarpour K (2019) Multi-grip classification-based prosthesis control with two EMG-IMU sensors. IEEE Trans Neural Syst Rehabil Eng 28:508–518. https://doi.org/10.1101/579367
57. Landgraf M, Yoo IS, Sessner J, Mooser M, Kaufmann D, Mattejat D, Reitelsh S (2018) Gesture recognition with sensor data fusion of two complementary sensing methods. In: 2018 7th IEEE international conference on biomedical robotics and biomechatronics (Biorob). IEEE, Enschede, Netherlands, pp 795–800. https://doi.org/10.1109/BIOROB.2018.8487949

58. Fougner A, Scheme E, Chan ADC, Englehart K, Stavdahl Ø (2011) Resolving the limb position effect in myoelectric pattern recognition. IEEE Trans Neural Syst Rehabil Eng 19:644–651. https://doi.org/10.1109/TNSRE.2011.2163529
59. Teramae T, Noda T, Morimoto J (2018) EMG-based model predictive control for physical human-robot interaction: application for assist-as-needed control. IEEE Robot Autom Lett 3:210–217. https://doi.org/10.1109/LRA.2017.2737478
60. Gui K, Tan U-X, Liu H, Zhang D (2020) Electromyography-driven progressive assist-as-needed control for lower limb exoskeleton. IEEE Trans Med Robot Bionics 2:50–58. https://doi.org/10.1109/tmrb.2020.2970222
61. Cisnal A, Alonso R, Turiel JP, Fraile JC, Lobo V, Moreno V (2019) EMG based bio-cooperative direct force control of an exoskeleton for hand rehabilitation: a preliminary study. In: Masia L, Micera S, Akay M, Pons J (eds) Converging clinical and engineering research on neurorehabilitation III. ICNR 2018. Biosystems & biorobotics. Springer, Cham, pp 390–394. https://doi.org/10.1007/978-3-030-01845-0_78
62. Scotto Di Luzio F, Simonetti D, Cordella F, Miccinilli S, Sterzi S, Draicchio F, Zollo L (2018) Bio-cooperative approach for the human-in-the-loop control of an end-effector rehabilitation robot. Front Neurorobot 12:1–12. https://doi.org/10.3389/fnbot.2018.00067
63. Novak D, Mihelj M, Ziherl J, Olenšek A, Munih M (2011) Psychophysiological measurements in a biocooperative feedback loop for upper extremity rehabilitation. IEEE Trans Neural Syst Rehabil Eng 19:400–410. https://doi.org/10.1109/TNSRE.2011.2160357
64. Mihelj M, Novak D, Munih M (2009) Emotion-aware system for upper extremity rehabilitation. In: 2009 Virtual rehabilitation international conference, VR 2009, pp 160–165. https://doi.org/10.1109/ICVR.2009.5174225
65. Guerrero CR, Fraile Marinero JC, Turiel JP, Muñoz V (2013) Using "human state aware" robots to enhance physical human-robot interaction in a cooperative scenario. Comput Methods Programs Biomed 112:250–259. https://doi.org/10.1016/j.cmpb.2013.02.003
66. Badesa FJ, Morales R, Garcia-Aracil N, Sabater JM, Casals A, Zollo L (2014) Auto-adaptive robot-aided therapy using machine learning techniques. Comput Methods Programs Biomed 116:123–130. https://doi.org/10.1016/j.cmpb.2013.09.011
67. Gümüslü E, Erol Barkana D, Köse H (2020) Emotion recognition using EEG and physiological data for robot-assisted rehabilitation systems. In: Gümüslü E, Barkana DE, Köse H (eds) Companion publication of the 2020 international conference on multimodal interaction. Association for Computing Machinery, New York, pp 379–387. https://doi.org/10.1145/3395035.3425199
68. Khezri M, Firoozabadi M, Sharafat AR (2015) Reliable emotion recognition system based on dynamic adaptive fusion of forehead biopotentials and physiological signals. Comput Methods Programs Biomed 122:149–164. https://doi.org/10.1016/j.cmpb.2015.07.006
69. Hariharan A, Adam MTP (2015) Blended emotion detection for decision support. IEEE Trans Hum Mach Syst 45:510–517. https://doi.org/10.1109/THMS.2015.2418231
70. Koelstra S, Mühl C, Soleymani M, Lee JS, Yazdani A, Ebrahimi T, Pun T, Nijholt A, Patras I (2012) DEAP: a database for emotion analysis; using physiological signals. IEEE Trans Affect Comput 3:18–31. https://doi.org/10.1109/T-AFFC.2011.15
71. Kim J, André E (2008) Emotion recognition based on physiological changes in music listening. IEEE Trans Pattern Anal Mach Intell 30:2067–2083. https://doi.org/10.1109/TPAMI.2008.26
72. Mandryk RL, Atkins MS (2007) A fuzzy physiological approach for continuously modeling emotion during interaction with play technologies. Int J Hum Comput Stud 65:329–347. https://doi.org/10.1016/j.ijhcs.2006.11.011
73. Liu C, Conn K, Sarkar N, Stone W (2008) Physiology-based affect recognition for computer-assisted intervention of children with autism spectrum disorder. Int J Hum Comput Stud 66:662–677. https://doi.org/10.1016/j.ijhsc.2008.04.003
74. Picard RW, Vyzas E, Healey J (2001) Toward machine emotional intelligence: analysis of affective physiological state. IEEE Trans Pattern Anal Mach Intell 23:1175–1191. https://doi.org/10.1109/34.954607

75. Zhai J, Barreto AB, Chin C, Li C (2005) Realization of stress detection using psychophysiological signals for improvement of human-computer interactions. In: Proceedings. IEEE SoutheastCon, 2005. IEEE, Ft. Lauderdale, FL, USA, pp 415–420. https://doi.org/10.1109/secon.2005.1423280
76. Konrad P (2005) The ABC of EMG: a practical introduction to kinesiological electromyography, 1st edn. Noraxon Inc., USA. https://doi.org/10.1016/j.jacc.2008.05.066
77. McManus L, de Vito G, Lowery MM (2020) Analysis and biophysics of surface EMG for physiotherapists and kinesiologists: toward a common language with rehabilitation engineers. Front Neurol 11. https://doi.org/10.3389/fneur.2020.576729
78. Bi L, Feleke A, Guan C (2019) A review on EMG-based motor intention prediction of continuous human upper limb motion for human-robot collaboration. Biomed Signal Process Control 51:113–127. https://doi.org/10.1016/j.bspc.2019.02.011
79. Fu J, Choudhury R, Hosseini SM, Simpson R, Park JH (2022) Myoelectric control systems for upper limb wearable robotic exoskeletons and exosuits—a systematic review. Sensors 22:8134. https://doi.org/10.3390/s22218134
80. Battista NA, Strickland WC, Miller LA (2017) IB2d: a python and MATLAB implementation of the immersed boundary method. Bioinspir Biomim 12. https://doi.org/10.1088/1748-3190/aa5e08
81. Hogan N, Krebs HI, Charnnarong J, Srikrishna P, Sharon A (1992) MIT—MANUS: a workstation for manual therapy and training I. In: 1992 Proceedings IEEE international workshop on robot and human communication (ROMAN), pp 161–165. https://doi.org/10.1109/ROMAN.1992.253895
82. Qiu S, Zhao H, Jiang N, Wang Z, Liu L, An Y, Zhao H, Miao X, Liu R, Fortino G (2022) Multi-sensor information fusion based on machine learning for real applications in human activity recognition: State-of-the-art and research challenges. Inf Fusion 80:241–265. https://doi.org/10.1016/j.inffus.2021.11.006
83. Harwin WS, Patton JL, Edgerton VR (2006) Challenges and opportunities for robot-mediated neurorehabilitation. Proc IEEE 94:1717–1726. https://doi.org/10.1109/JPROC.2006.880671
84. Cisnal A, Antolínez D, Turiel JP, Carlos Fraile J, De La Fuente E (2023) A versatile embedded platform for implementation of biocooperative control in upper-limb neuromotor rehabilitation scenarios. IEEE Access 11:35726–35736. https://doi.org/10.1109/ACCESS.2023.3265898

Chapter 2
Hypotheses and Objectives

The pursuit of developing cost-effective and reliable biocooperative systems for neuromotor rehabilitation, with the aim of improving the quality of life for individuals with motor disabilities, has emerged as a significant concern in recent years. Consequently, the proposal developed in this Doctoral Thesis has been directed towards the creation of low-cost embedded systems for physiological data acquisition and the implementation of reliable control paradigms. The primary objective of these systems is to enhance the accessibility and integration of biocooperative technologies into real clinical practices. To provide a comprehensive framework for this thesis, the hypotheses that have served as the basis for each study, as well as the overarching hypothesis that justifies the present Doctoral Thesis, are explicitly declared in Sect. 2.1. Additionally, Sect. 2.2 outlines the principal objective and the specific sub-objectives that must be fulfilled in order to accomplish this overarching goal.

2.1 Hypothesis

Despite the increasing interest in biocooperative systems for neuromotor rehabilitation in scientific literature, their current limitations have confined their application primarily to laboratory settings. As a result, studies often overlook their potential utilization in real-world scenarios and focus solely on academic purposes. Thus, a naïve hypothesis can be formulated: *limitations of current biocooperative control systems that restrict their application outside laboratory settings can be mitigate.* However, while this statement serves as a starting point, it is not comprehensive enough to address a specific research question. Therefore, further exploration and examination of lower-level hypotheses are necessary to advance toward this goal.

As stated in subsection 1.4, current biocooperative systems encounter two primary limitations: (I) hardware, and (2) reliability. Regarding the hardware limitations

A. Cisnal de la Rica, *Development of Biocooperative Control Strategies for Neuromotor Rehabilitation Robotic Platforms*, Springer Theses,
https://doi.org/10.1007/978-3-032-02485-5_2

(I) improving user acceptance requires a compact set of appropriately positioned sensors, thereby necessitating size reduction, even if it imposes limitations on battery capacity [1]. Additionally, the development of integrated multimodal approaches is crucial for user-centric systems in rehabilitation applications, fostering engagement within the user community [2]. Real-time embedded systems play a vital role in reducing system response time, as lengthy delays can have detrimental effects, such as decreased task completion accuracy, reduced perceptual sensitivity, and increased task error rates, all of which can negatively impact user acceptance and satisfaction [3].

By developing small multimodal platforms with high computational capacities, user acceptance can be improved in terms of comfort and faster system response. Furthermore, the creation of affordable and versatile platforms that support real-time multimodal controls would encourage further research in the field of biocooperative controls. With improved accessibility to this technology and continued research, it would be possible for researchers to effectively translate biocooperative control strategies that rely on physiological signals into practical and affordable real-world solutions.

The limitation of reliability (II) largely arises as a result of hardware limitations. By enhancing accessibility, research can be increased, leading to the development of robust and effective controls that can be implemented in the clinical setting. Moreover, when considering the specific case of myoelectric EMG-driven controls, which are the most common in the field of robotic rehabilitation, their implementation in real-world applications is not viable due to reliability issues [4]. Although various sEMG-based pattern recognition methods have shown promising results, their practical application is limited due to the need for a large number of DoFs in the robot and the associated high computational costs, making them unsuitable for real-time embedded systems.

The development of EMG-driven control systems should focus on providing controls that can be integrated into existing robotic platforms with a limited number of DoFs. This approach reduces the number of electrodes, improving both user and clinician acceptance by reducing the time required for electrode placement. Moreover, a lower number of DoFs and restricted movements contribute to increased system reliability, robustness, and decreased computational costs, resulting in appropriate response times. Collectively, these factors enhance user acceptance. These statements form the core of the present Doctoral Thesis, and they can be combined into the following:

> Biocooperative systems may be oriented toward a real use outside laboratory by developing cost-effective, real-time, embedded physiological data acquisition systems, which promote further research in developing reliable and robust controls.

2.2 Objectives

The overall goal of this Doctoral Thesis was to design, develop and evaluate biocooperative control strategies in the context of upper-limb neuromotor rehabilitation, and to provide affordable technology for their implementation in an attempt towards widespread use in clinical settings. This general objective involved the design and development of embedded systems for physiological signal acquisitions and its integration in neuromotor rehabilitation systems by implementing and evaluating assistive control strategies. In order to achieve this general objective, the following specific objectives arise:

I. Conduct a comprehensive literature review and examine the current state-of-the-art in upper-limb neuromotor rehabilitation platforms, with particular emphasis on control strategies that integrate the human into the control loop through the analysis of physiological signals.
II. Design and development affordable embedded systems for physiological signal recording and data processing for real-time execution of the control paradigm of the rehabilitation systems.
III. Design and development of biocooperative control strategies for the developed real-time embedded solutions and their integration into pre-existing rehabilitation platforms.
IV. Perform a comprehensive evaluation of the performance of the proposed biocooperative rehabilitation systems, including the assessment of control accuracy, time response, and user performance.
V. Evaluation of the performance of the wearable multimodal embedded acquisition system, focusing on its versatility, power consumption and reliability of the recorded signals for the implementation of biocooperative controls.

2.3 Thesis Organization

This thesis is structured as a compendium of three scientific publications, each addressing a key aspect of the development of low-cost physiological acquisition systems and biocooperative control strategies for upper-limb robotic rehabilitation. This area is of particular clinical relevance, as upper-limb paresis is one of the most common consequences of stroke and significantly impairs patients' independence and quality of life.

This book has been written based on the research findings and results reported in these three publications. Each publication corresponds to a distinct phase of the research. The first and second studies focus on EMG-driven neuromotor rehabilitation using a robotic hand exoskeleton and a custom-designed, low-cost real-time embedded EMG acquisition system. The third presents a highly versatile, wearable, and cost-effective real-time embedded platform for implementing multimodal biocooperative controls.

The first article [5] presents the design of a low-cost 2-channel EMG real-time embedded solution and its integration in a hand rehabilitation exoskeleton. A non-pattern recognition, threshold-based bilateral control strategy was developed, enabling the detection of gestures from the unimpaired hand and replicating them on the exoskeleton worn on the paretic hand. The system was evaluated in terms of classification, accuracy and response time, yielding satisfactory performance.

Building on these results, the second article [6] investigates the impact of EMG-based visual feedback on user performance. The study demonstrated that visualizing muscle activation in real time allowed users to gain better control over the exoskeleton's movement, adjust their exerted effort, and improve their ability to self-regulate EMG responses.

Following the validation of the EMG system, the third study [7] expands the solution to include additional physiological signals relevant to biocooperative upper-limb rehabilitation. The resulting multimodal embedded system integrates inertial measurement units (IMU), electrocardiography (ECG), electromyography (EMG), galvanic skin response (GSR), and skin temperature (SKT) sensors. The system was evaluated in two representative rehabilitation scenarios: (1) a VR-based upper-limb exergame using EMG and IMU tracking, and (2) an AAN wrist rehabilitation protocol that modulated assistance based on the user's physiological state derived from GSR, ECG, and SKT. In both cases, the system delivered high-quality signals, robust processing performance, and adequate battery life.

Title and the citation of each of the papers of the present doctoral thesis are shown below:

RobHand: A Hand Exoskeleton with Real-Time EMG-Driven Embedded Control. Quantifying Hand Gesture Recognition Delays for Bilateral Rehabilitation [5]
A. Cisnal, J. Pérez-Turiel, J.C. Fraile, D. Sierra and E. de la Fuente, "RobHand: A Hand Exoskeleton with Real-Time EMG-Driven Embedded Control. Quantifying Hand Gesture Recognition Delays for Bilateral Rehabilitation," in IEEE Access, vol. 9, pp. 137,809–137,823, 2021, https://doi.org/10.1109/ACCESS.2021.3118281.

Interaction with a Hand Rehabilitation Exoskeleton in EMG-Driven Bilateral Therapy. Influence of Visual Biofeedback on the Users' Performance [6]
A. Cisnal, P. Gordaliza, J. Pérez-Turiel and J.C. Fraile, "Interaction with a Hand Rehabilitation Exoskeleton in EMG-Driven Bilateral Therapy: Influence of Visual Biofeedback on the Users' Performance," in Sensors, vol. 23, 2048, 2023, https://doi.org/10.3390/s23042048.

A Versatile Embedded Platform for the Implementation of Biocooperative Controls in Upper-limb Neuromotor Rehabilitation Scenarios [7]
A. Cisnal, D. Antolínez, J. P. Turiel, J. C. Fraile and E. De La Fuente, "A Versatile Embedded Platform for Implementation of Biocooperative Control in Upper-Limb Neuromotor Rehabilitation Scenarios," in IEEE Access, vol. 11, pp. 35,726–35,736, 2023, https://doi.org/10.1109/ACCESS.2023.3265898.

References

1. Qiu S, Zhao H, Jiang N, Wang Z, Liu L, An Y, Zhao H, Miao X, Liu R, Fortino G (2022) Multi-sensor information fusion based on machine learning for real applications in human activity recognition: state-of-the-art and research challenges. Inf Fusion 80:241–265. https://doi.org/10.1016/j.inffus.2021.11.006
2. Rodgers MM, Alon G, Pai VM, Conroy RS (2019) Wearable technologies for active living and rehabilitation: current research challenges and future opportunities. J Rehabil Assist Technol Eng 6:1–9. https://doi.org/10.1177/2055668319839607
3. Yang E, Dorneich MC (2015) The effect of time delay on emotion, arousal, and satisfaction in human-robot interaction. In: Proceedings of the human factors and ergonomics society 59th annual meeting 2015, pp 443–447. https://doi.org/10.1177/1541931215591094
4. Meattini R, Benatti S, Scarcia U, De Gregorio D, Benini L, Melchiorri C (2018) An sEMG-based human-robot interface for robotic hands using machine learning and synergies. IEEE Trans Compon Packag Manuf Technol 8:1149–1158. https://doi.org/10.1109/TCPMT.2018.2799987
5. Cisnal A, Perez-Turiel J, Fraile JC, Sierra D, De La Fuente E (2021) RobHand: a hand exoskeleton with real-time EMG-driven embedded control. Quantifying hand gesture recognition delays for bilateral rehabilitation. IEEE Access 9:137809–137823. https://doi.org/10.1109/ACCESS.2021.3118281
6. Cisnal A, Gordaliza P, Pérez Turiel J, Fraile JC (2023b) Interaction with a hand rehabilitation exoskeleton in EMG-driven bilateral therapy: influence of visual biofeedback on the users' performance. Sensors 23:2048. https://doi.org/10.3390/s23042048
7. Cisnal A, Antolínez D, Turiel JP, Carlos Fraile J, De La Fuente E (2023a) A versatile embedded platform for implementation of biocooperative control in upper-limb neuromotor rehabilitation scenarios. IEEE Access 11:35726–35736. https://doi.org/10.1109/ACCESS.2023.3265898

Chapter 3
Materials and Methods

This chapter describes the materials and methods that were applied throughout the compendium of publications. Section 3.1 details the robotic platforms used within the scope of this study. Section 3.2 is dedicated to a comprehensive review of the relevant literature pertaining to biocooperative controls. Subsequently, Sect. 3.3 focuses on experimental designs and setups. Finally, Sect. 3.4 offers a description of the applied metrics for performance assessment.

3.1 Robotic Rehabilitation Platforms

Two robotic platforms were used in this Doctoral Thesis: the RobHand and the M3Rob. The RobHand was used in the first [1] and second articles [2], in combination with the designed EMG acquisition system. On the other hand, the M3Rob was employed in the last article [3] to assess the performance of the wearable multimodal acquisition system. The electromechanical structures of these robotic platforms offer unique features and capabilities, enabling the investigation of different biocooperative controls. In this section, an overview of each platform's specifications is provided to help readers understand the technology behind them.

3.1.1 RobHand, a Robot for Hand Rehabilitation

The RobHand (Robot for Hand Rehabilitation) project aimed to develop a robot for hand rehabilitation for patients with neuromotor disabilities resulting from a cerebrovascular accident. The project was developed in partnership between the TICCYL Digital S.L. enterprise and the ITAP Robotics Research Group at the School of Industrial Engineering of the University of Valladolid. The work was supported by

A. Cisnal de la Rica, *Development of Biocooperative Control Strategies for Neuromotor Rehabilitation Robotic Platforms*, Springer Theses,
https://doi.org/10.1007/978-3-032-02485-5_3

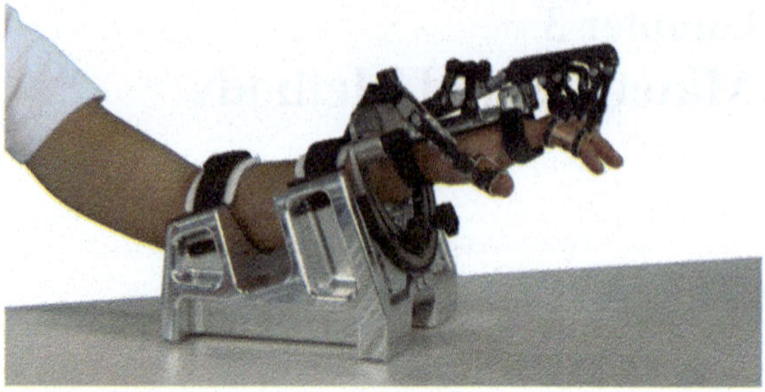

Fig. 3.1 The RobHand robotic platform and forearm support for hand rehabilitation [1]

CDTI (Center of Development of Industrial Technology), a Public Business Entity of the Spanish Ministry of Science and Innovation, under project IDI-20170263 with the European Regional Development Fund (ERDF).

RobHand is an exoskeleton-type robot for neuromotor rehabilitation, which assists flexion and extension of the hand fingers (Fig. 3.1). It was specifically designed to be lightweight, easy-to-use, cost effective and versatile. Its mechanical design is based on a direct-driven under-actuated serial-bar linkage paradigm. In particular, the mechanical structure is based on a platform located on the back of the hand and five subassemblies, each one associated to one finger. All of them are mounted on the platform except the one for the thumb, which is connected with the platform using a linkage device. The subassemblies are composed of an underactuated linkage-rotate mechanic and a flexible double-ring to transmit the linear force from the actuator to the ring attached to the proximal and medial phalanges [4].

Thus, the device is characterized by independent finger motion and has two DoFs for each finger: one active DoF associated with the metacarpophalangeal (MCP) joint and passive DoF associated with the proximal interphalangeal (PIP) joint. The robot provides a ROM of 8° of extension and 62° of flexion in the MCP joint. The exoskeleton incorporates five L12-30-100-6-I linear actuators (Actuonix Motion Devices Inc., Saanichton, BC, Canada). The actuators have a 30 mm stroke, provided up to 23 N force and are low-cost, reducing the overall cost of the system and making it more affordable.

Additionally, the flexible double-rings ensure an easy donning and doffing of the exoskeleton even in case of patients suffering from severe hand spasticity. First, the rings are placed on the fingers and then, they are joined to their corresponding linkage-rotate mechanisms. The exoskeleton is adjusted to the hand by using two Velcro straps. Additionally, the thumb submodule is manually adjustable, providing an easy adaptation to different hand sizes. The system is built in aluminum except for the linkage-rotate mechanisms and the flexible double-rings, which are 3D printed, resulting in a total weight of 610 g. The system also incorporates a forearm support to compensate forces created by the exoskeleton weight.

3.1.2 M3Rob, a Robot for Wrist Rehabilitation

The M3Rob (in Spanish, *Mente-Mano-Muñeca*, Mind-Hand-Wrist-Robot) project aimed to develop a platform for neuromotor and cognitive hand and wrist rehabilitation for people who have suffered a cerebrovascular accident. The project is developed by TICCYL Digital S.L. company, the ITAP Robotics Research Group and the Biomedical Engineering Group, both from the University of Valladolid, and the *Benito Menni* Hospital Center in Valladolid. The project is supported by the Ministry of Science and Innovation of Spain, under grant RTC2019-007350-1.

The rehabilitation robot assists the pronation/supination (PS), flexion/extension (FE), and radial/ulnar (RU) deviation motions of the wrist. It is based on a three-serial revolute active joint (3-DoF RRR) mechanism (Fig. 3.2). Each rotational joint is powered by an independent brushed DC motor (Maxon Motors, Switzerland) equipped with a 3-channel encoder. Cables are used to transmit the force generated by the motors to the mechanical structure, and thus, assist the wrist motion. The mechanism also incorporates one passive DoF that allows manual adjustment of the forearm support. A 6-axis torque-force sensor K6D27 50N/1Nm (Me&Systeme, Germany) is embedded in a cylindrical handle. The range of motion of PS, FE and RU joints are 180°, 135° and 110°, respectively [5, 6].

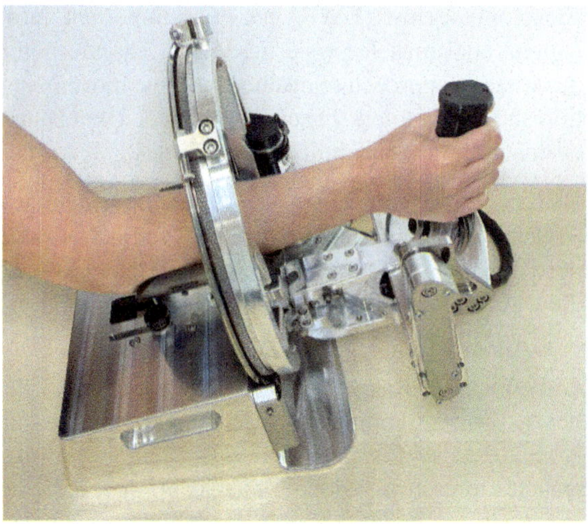

Fig. 3.2 The M3Rob robotic platform for wrist rehabilitation, which provides assistance of the pronation/supination, flexion/extension, and radial/ulnar deviation motions

3.2 Biocooperative Controls

Section 3.2.1 describes some physiological signals, which are necessary for the development of biocooperative controls. Subsequently, Sect. 3.2.2 focuses on reviewing relevant literature on non-pattern recognition-based myoelectric controls. EMG-driven control can be classified into two main types: discrete and continuous control, with the former further categorized into non-pattern (or threshold) methods and pattern recognition methods. Considering our emphasis on developing an embedded control system with intrinsic limited computation capacity and intended for real applications, we have opted to concentrate only on non-pattern recognition controls. Finally, Sect. 3.2.3 presents a brief summary of the state of the art on multimodal controls will be presented.

3.2.1 Physiological Signals

Based on a wide range of physiological signals used in related studies (Table 1.2) and considering both usability and wearability, we decided to only focus on four sensor technologies from which valuable physiological data can be extracted:

Electromyogram (EMG) provides important information on muscle activation patterns and muscle properties. EMG has the potential to offer an objective and quantitative approach to evaluate not just movement patterns, but also muscle function and local fatigue muscle [7]. Hence, it can be used for progress assessment and evaluation of the rehabilitation outcomes.

Electrocardiogram (ECG) provides information about the activity of the rate. Several parameters can be extracted, such as HR or HRV. HR can indicate the body's need for oxygen and provide insights into physical effort. Additionally, HR and arousal are strongly correlated [8]. HRV as well as HR has been also used as an indicator of arousal, physical effort and mental workload [9].

Galvanic Skin Response (GSR) signal can be divided into tonic and phasic components. The tonic component, also known as SCL, exhibits slow changes over time and is related to skin hydration, dryness, and autonomic regulation, indicating general changes in arousal. On the other hand, the phasic component, referred to as SCR, represents the rapidly changing part of the signal that responds to emotionally stimulating events. SCL a is a good indicator for physical workload, while the frequency of SCR increases with arousal and mental workload [10].

Peripheral Skin Temperature (SKT) is an effective way to estimate emotional state [11]. When the sympathetic nervous system is activated due to stress, it causes vasoconstriction, which reduces peripheral circulation and lowers skin temperature. SKT changes can also indicate significant physical activity as the shunting of blood in the body helps regulate temperature and meet the oxygen demands of organs.

3.2.2 Non-pattern EMG-Based Controls

Non-pattern EMG-based controls require the comparison of EMG activity between muscles. To achieve this, signal normalization is necessary to mitigate the effects of interferences on the signals, including power line noise, skin perspiration, sensor contact impedance, and crosstalk interference from active muscles [12]. These factors contribute to high signal variability, and small differences in electrode placement between sessions can compromise the repeatability of EMG signals [13]. EMG processing involves rectifying the raw EMG signal and mapping it into a normalized signal ranging from 0 to 1.

EMG rectification involves converting all negative signal values to positive amplitudes. The average rectified value (AVR), root mean squares (RMS) and signal envelope are commonly used methods for full-wave signal rectification, typically followed by a low-pass filtering [14]. However, while RMS measures the signal power, AVR and envelope do not possess a specific physical meaning [15].

Normalization typically involves using the maximal voluntary contraction (MVC) as a reference value. This approach allows for the comparison of muscle activation patterns within an individual over time. However, it is not suitable for comparing between different individuals, muscles, or tasks. In clinical applications, where patients may not be capable of exerting maximum efforts or repeating certain movements, EMG is often normalized to submaximal contractions or reference voluntary contractions (RVC) as they provide a more reliable indicator of changes in muscular activity [16].

Various non-pattern recognition EMG-based controls for hand robots have been previously proposed in the literature. These controls share a similar nature but are referred to differently, such as binary control [17], ON–OFF control [18], time-over-threshold control [19], or triggered mode [20, 21].

A binary control was developed to regulate the pressure level of a pneumatic hand exoskeleton [17]. The biceps EMG signal was recorded, rectified, smoothed using a Butterworth low-pass filter, and normalized using the MVC. The resulting EMG signal was then used to determine the controller binary output: if the EMG signal exceeded a specified threshold value, the output was set to "on", and if it was below the threshold, the output was set to "off". To avoid output oscillation, a hysteresis mechanism was implemented in the valve triggering system. Initially, the threshold values were set at 55% MVC to activate the "on" state and 45% MVC to initiate the "off" state. However, before each experiment, these threshold values were adjusted to ensure they corresponded to the subject's comfortable level.

A triggered ON–OFF EMG-driven control for the GLOREHA glove was designed to identify three different states, namely hand opening, hand rest and hand closing, with a predefined speed command [18]. Electrodes were placed on the extensor digitorum (ED) and the extensor carpi radialis to detect muscle activation during hand opening, while the palmaris longus and the flexor carpi ulnaris muscles were used to detect hand closing. The EMG signals were sampled at a rate of 25 kHz, and a temporal segmentation of 1 s windows with a 99.5% overlap between adjacent windows was employed, resulting in a temporal resolution of 5 ms. Within each window, the signal was rectified using RMS method. The difference between the rectified signals of the extensor and flexor EMG signals was then calculated. This difference, along with two threshold values, was employed to determine the three states.

Another EMG-triggered control utilized two EMG signals obtained from the ED and the abductor pollicis brevis (APB) muscles to detect the intention of hand opening and closing [20]. The EMG signals were subjected to full-wave rectification and then smoothed by applying a moving average filter with a window size of 100 ms, followed by normalization using the MVC. The hand rehabilitation robot was programmed to initiate movement when the EMG signal exceeded a threshold of 30% of the MVC.

An open-loop time-over-threshold EMG-driven control was designed for a hydraulically actuated soft rehabilitation glove [19]. It detected user intention by measuring electromyographic signals from electrodes attached to the forearm, specifically the flexor digitorum superficialis (FDS) and the ED. The control system continuously monitors and compares the FDS and EDC muscle signals to detect predefined conditions: flex, extend, and hold. The flex condition pressurizes the soft actuators, causing the glove to flex along with the fingers, while the extend condition depressurizes the actuators, returning the fingers to the extended position. The hold condition maintains the present fluidic pressure within the actuators. To avoid misinterpretation of involuntary muscle contractions, the conditions require the processed signals to cross predefined thresholds for a specified duration. These conditions are manually adjusted.

An EMG-triggered mode strategy was developed to control the Hand of Hope exoskeleton by recording the EMG signals from the APB and ED muscles [21] The EMG signals were normalized with respect to the MVC, which was determined at the beginning of each training session. A threshold of 20% of the MVC was employed to initiate hand opening and closing motions. During hand closing triggering mode, the robotic system awaited EMG signals from the APB muscle surpassing 20% MVC before initiating hand closure. Similarly, during hand opening mode, the system awaited EMG signals from the ED muscle exceeding 20% MVC before starting the hand opening action.

3.2.3 Multimodal Controls

Multimodal controls and methods for recognizing emotions through physiological signals have been previously proposed in the literature. One study adapted the level of assistance provided by an upper-limb rehabilitation robot based on the patient's performance and fatigue. Patient performance was determined by using biomechanical information from an IMU sensor, while muscular fatigue was estimated using EMG signals [22]. Another study adjusted the difficulty of an upper extremity rehabilitation task using biomechanics (force and movement), task performance, and physiological signals, such as ECG, GSR, RESP, and SKT [23].

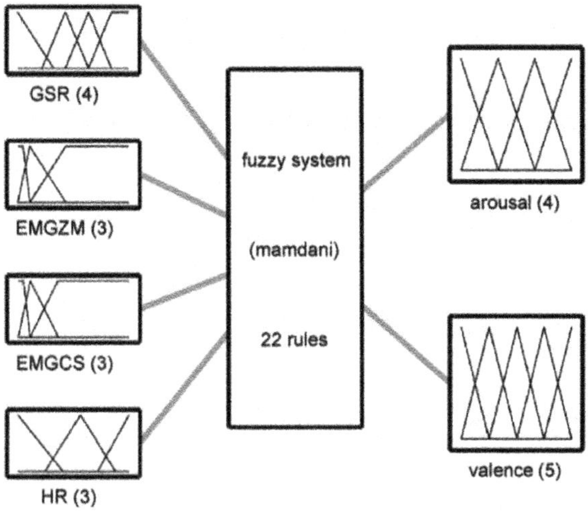

Fig. 3.3 Determination of arousal and valence from GSR, EMG and HR information [26]

A two-stage fuzzy logic model was used to generate the next action primitives of an upper extremity rehabilitation device. The first stage calculated motor performance, arousal, and valence based on position, force, RESP, GSR, and SKT. The second stage selected action primitives based on physical effort associated with motor performance, arousal, and valence [24].

Affective and emotional states have also been used to control devices, such as modulating the assistance provided by a haptically controlled robot based on user emotions. Emotions were estimated by a fuzzy logic model that considered the 3-dimensional emotion model (arousal, dominance, and valence) using HR mean, SCL mean, and SCR frequency as inputs [25]. Similarly, arousal and valence have been determined using normalized GSR, HR, and EMG signals as inputs to a fuzzy logic control scheme characterized by 22 rules as shown in Fig. 3.3 [26].

Emotion recognition, considering either pleasant, neutral, or unpleasant states, has been addressed using EEG, BVP, SKT, and SCL recordings and developing gradient boosting machines (GBMs) and convolutional neural networks (CNNs) [27]. Models were developed to recognize six and even eight basic emotions employing various classification methods such as SVM and k-nearest neighbors (KNN) [28, 29]. Detection of basic emotions considering two, three, or four states was also investigated using HR and SCL measurements and employing a Classification and Regression Trees (CART) algorithm [30]. Several studies have used physiological signals to map emotions in a two-dimensional model of arousal and valence [31, 32].

Affection recognition has been developed considering three target affective states, such as anxiety, engagement, and liking, using SVM-based recognition models with different features derived from physiological signals [33]. Differentiating stress and normal states has been carried out using SVM and features derived from BVP, GSR, and PD signals. Finally, a classification method was developed to detect three levels of stress using pulse rate, RESP rate, SKT, and GSR features [34].

3.3 Experimental Design

This section details the experimental design employed to investigate the accuracy and responsiveness of EMG-driven control (Sect. 3.3.1) and the impact of the EMG-based visual feedback on user performance (Sect. 3.3.2).

3.3.1 *Accuracy and Responsiveness of the EMG-Driven Control*

The first study was conducted to evaluate the accuracy and responsiveness of the EMG-driven bilateral assistance with the RobHand exoskeleton [1]. Ten healthy subjects (7 males, 3 females) over 18 years old with no neurological or motor impairments, volunteered for the study and provided written informed consent.

The subjects wore a 5DT Data Glove (5DT Technologies) to measure the actual position of the dominant hand (corresponding to the non-paretic hand of the patient), while the sEMG signals of the muscles responsible for that hand movement were recorded, along with the recognized hand gesture. The subjects wore the hand exoskeleton on the non-dominant hand (corresponding to the paretic hand of the patient), which replicated the gestures recognized from the sEMG signals analysis (Fig. 3.4).

The subjects were asked to perform an initial calibration followed by ten one-minute tests, with a 5 min rest between tests to avoid the appearance of muscle fatigue. Each test consisted of performing and maintaining hand gestures (rest, open and close) with the dominant hand. The gesture to be performed was indicated on the computer screen and was randomly generated every 5 s.

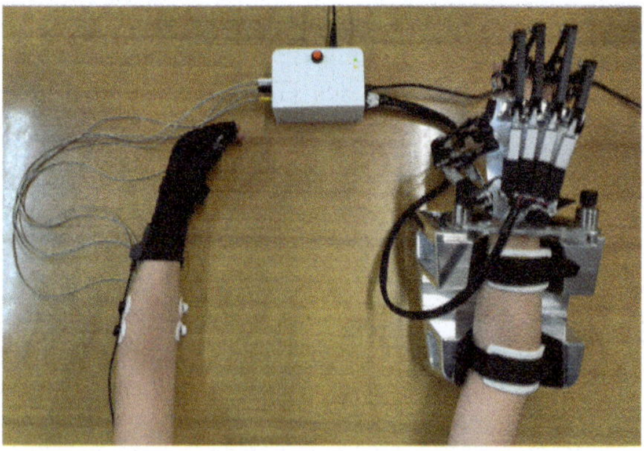

Fig. 3.4 The left-handed subject wearing the 5DT Data Glove on the dominant hand and the exoskeleton on the non-dominant hand. The sEMG electrodes are attached to the target muscles of the forearm of the dominant hand [1]

3.3.2 Influence of the Feedback on the User's Performance

The second study was conducted to evaluate the influence of EMG-based visual feedback on the subject's performance when performing EMG-driven bilateral therapies with the RobHand exoskeleton [2, 3]. Eighteen subjects (23 ± 3.4 years old) with no neurological or motor impairments volunteered and provided written informed consent.

The subjects wore the hand exoskeleton on the non-dominant hand (in the case of a patient, it would correspond to the paretic hand), while the sEMG signals were recorded from the target muscles of the dominant hand (in the case of a patient, it would correspond to the healthy hand) (Fig. 3.5).

Subjects performed an initial calibration and four one-minute experimental tests (named A, B, C and D) with a three-minute break between tests to avoid muscular fatigue. To avoid the appearance of a learning order effect, the four tests were randomly performed. Each test consisted of performing and maintaining a hand gesture (rest, open or close) with the dominant hand. The target gestures were randomly generated every three seconds. Each test was characterized by a different combination of two feedback sources: kinesthetic and EMG-based visual feedback (Table 3.1). While the kinesthetic feedback was provided by the movement of the hand exoskeleton based on the EMG-driven bilateral control, the EMG-based visual feedback was provided by the computer screen.

The overall system configuration employed in the experimental protocol is depicted in Fig. 3.6. The participant is instructed to execute a specific hand gesture (i.e., open, close, or rest) using their dominant hand, guided by visual and auditory information provided by the computer. Concurrently, the recorded EMG signals are transmitted to the microcontroller for gesture recognition and generation of control signals to actuate the exoskeleton accordingly. The recognized gestures are then transmitted to the PC to update the EMG-based visual feedback. Additionally, both

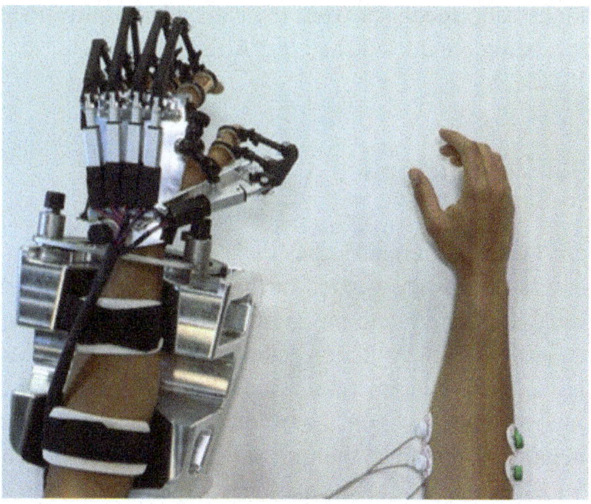

Fig. 3.5 The right-handed subject wearing the hand exoskeleton on the non-dominant hand while the sEMG electrodes are attached to the target muscles of the forearm of the dominant hand [2, 3]

Table. 3.1 Test configuration for evaluating user performance based on the presence of feedback sources, including kinesthetic and EMG-based visual feedback

	Kinesthetic feedback	EMG-based visual feedback
Test A	(✓)	(✓)
Test B	(✓)	(✗)
Test C	(✗)	(✓)
Test D	(✗)	(✗)

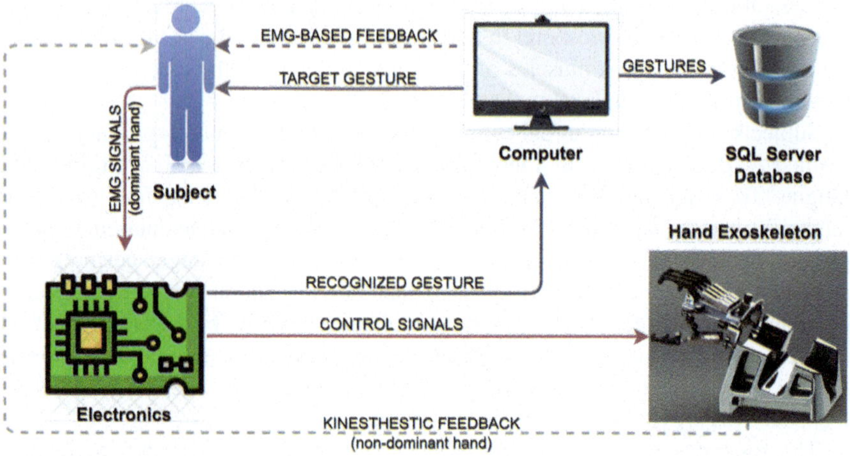

Fig. 3.6 Experimental setup diagram, showing the data flow between subsystems: visual information from the computer (yellow lines), EMG and control signals (red lines), data transmission (blue lines), and exoskeleton movement (green lines). The presence of the source of feedback (dotted lines) varies depending on the specific test [2]

the recognized and target gestures are stored in a SQL database as temporal series for subsequent offline analysis. The provision of EMG-based visual feedback and kinesthetic feedback from the exoskeleton movement is dependent on the specific test being conducted (i.e., test A, B, C, or D).

3.4 Performance Assessment

During the different studies that compose the present Doctoral Thesis, performance assessments were carried out to evaluate the real-time control performance of the rehabilitation system based on motion intention recognition. In the specific case of an assistive rehabilitation robot based on gesture recognition, a reliable and natural human–robot interaction requires a combination of good responsiveness and accurate gesture detection. Large delay times can exert adverse consequences and can negatively influence the user satisfaction towards robot-assisted therapy [35]. Likewise, wrong movement assistance negatively affects satisfaction and can even lead to physical damage. Hence, determining the relevant time delays and gesture classification metrics of the robotic system for an efficient HRI is essential [1]. Additionally, similarity measurements for time series and parametric statistical tests are used to evaluate user performance under different feedback setup conditions when using the EMG-driven hand exoskeleton [2].

3.4.1 Classification Accuracy

Evaluation metrics play a critical role in evaluating the effectiveness of classification problems. There are two types of classification problems according to the number of classes: binary classification and multi-class classification [36]. The quality of the classifier can be expressed using a confusion matrix, which is a table that records the number of occurrences between two raters: the rows of the table represent the predicted classes, while the columns show the actual/true classes.

For binary classification, the confusion matrix is shown in Fig. 3.7. True Positive (TP) and True Negative (TN) are the number of positive and negative classes that are correctly classified. Meanwhile, the numbers of misclassified negative and positive instances are denoted by False Positive (FP) and False Negative (FN), respectively [37].

From the confusion matrix, many metrics for evaluating the classification effectiveness can be calculated [38]. Accuracy (Acc) is the most popular classification, although some complement metrics such as precision or recall, are also used to express all the relevant information about algorithm performance. Accuracy measures how much the algorithm is correctly predicting in the entire set of data by evaluating the percentage of correct predictions over the total number of samples. Accuracy is calculated using Eq. (3.1), which considers the sum of TP and TN

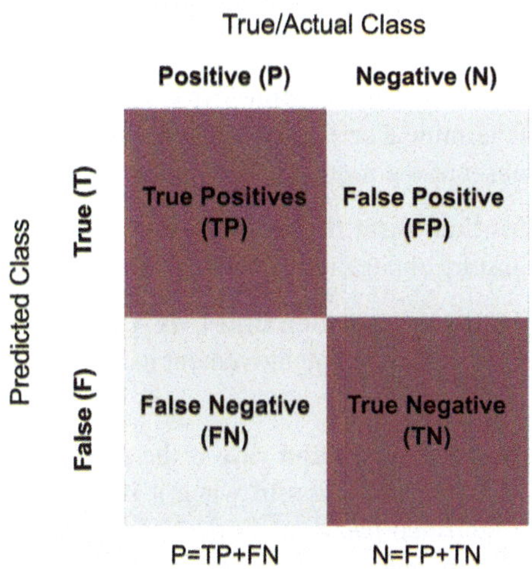

Fig. 3.7 Confusion matrix for a binary classification problem. There are two true classes: positive (P) and negative (N). The output of the predicted class is either true (T) or false (F)

elements at the numerator and the sum of all entries at the dominator.

$$Accuracy = \frac{TP + TN}{TP + FP + FN + TN} \quad (3.1)$$

The precision, or positive predictive value (PPV), quantifies the correctly predicted positive classes among the total predicted positive classes. Precision is calculated using Eq. (3.2).

$$Precision = \frac{TP}{TP + FP} \quad (3.2)$$

Recall measures the fraction of positive classes that are correctly classified. It is also known as true positive rate (TPR), and it is calculated using Eq. (3.3).

$$Recall = \frac{TP}{TP + FN} \quad (3.3)$$

The EMG-based gesture recognition control is considered a classification problem since it requires determining one gesture out of three possible gestures (open, close and rest). Hence, the confusion matrix and abovementioned metrics are used to evaluate the performance of the gesture recognition algorithm [1].

3.4.2 Time Delay Analysis

The study of responsiveness of the EMG-based rehabilitation system is carried out using the following time-related metrics [1]. Some of them were proposed by Li. et al. in the work in which they quantified the time delay of an EMG-based pattern recognition control of a virtual arm [39], the rest are proposed in the context of this work.

Motion-selection time (MST) is the time needed by the controller to accurately determine a certain gesture. Hence, MST is calculated as the time interval from the onset of the motion to the instant the controller accurately predicts that motion.

Motion-onset time (MOT) is calculated as the time interval from the onset of the gesture change movement to the instant the assistive robot starts that movement.

Motion-completion time (MCT) is calculated as the time interval from the onset of the gesture change movement to the instant the assistive robot correctly reaches the next gesture.

Motion-completion rate is the percentage of successful motions, considering that a motion is successful when it is performed within a time limit. The time limit is established based on clinical experience and is 5 s to ensure that the movement is not too slow so as not to demotivate or annoy the user.

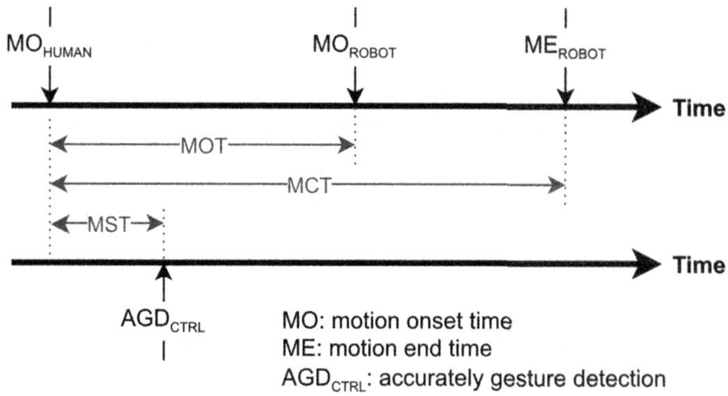

Fig. 3.8 Timing diagram showing the time-related metrics: MST, MOT, MCT [1]

A time diagram representing the time parameters (MST, MOT and MCT) is shown in Fig. 3.8 to enhance comprehension. These time-delay metrics are characterized by events that define their onset and end time: motion onset time of the subject (MO_{Human}), the motion onset time of the assistive robot (MO_{ROBOT}), the motion end time of the assistive robot (ME_{ROBOT}), and the time in which the controller accurately detects the gesture (AGD_{CTRL}).

3.4.3 Time Series Similarity Measurement

A similarity measure is mathematically defined as a real-valued function that quantified the similarity between two entities, in this context between two time series. Similarity measures can be divided into lock-step measures that calculate the one-to-one point Lp-norm or distance between two series or using elastic measurements that do not considered a fixed step, such as the Dynamic Time Warping (DTW) or the Longest Common Subsequence (LCSS) [40].

Lock-step measures are commonly used since they are relatively straightforward, intuitive and their low linear computational cost. However, since distance is computed from ith point of one time series to the ith point of another series, this method is highly sensitive to noise and time misalignments [40]. Therefore, two time series must be time-synchronized before calculating the distance measurements, since local time shifting can't be handled.

Synchronization of two time series x and y can be performed by calculating the lag at which their cross-correlation (r_{xh}) is the highest. r_{xh} is computed based on Eq. (3.4), where h is the lag and * denote the complex conjugate.

$$r_{xy}(h) = \begin{cases} \sum_{n=0}^{N-h-1} x(n+h) \cdot y^*(n), 0 \le h \le N-1 \\ r_{yx}^*(h), \quad -(N-1) \le h \le 0 \end{cases} \tag{3.4}$$

The most typical examples of distance metrics are the Manhattan distance (L1 norm) and the Euclidean distance (L2 norm). L1 norm is calculated as the sum of the absolute difference of the values of the time series as shown in Eq. (3.5), while L2 norm is calculated as the square root of the sum of the square time series values as expressed in Eq. (3.6).

$$L_1(x, y) = \sum_{i=0}^{n_{samples}} |x_i - y_i| \tag{3.5}$$

$$L_2(x, y) = \sqrt{\sum_{i=0}^{n_{samples}} (x_i - y_i)^2} \tag{3.6}$$

were x_i and y_i are elements of x and y, respectively. The comparison of distance measurements between time series with different durations requires a previous normalization. These metrics can be normalized by multiplying by the sampling period (T_s).

3.4.4 Statistical Analysis

Statistical analysis is a powerful tool in scientific research, allowing for fair comparisons of data across multiple groups and providing valuable insights into the relationship between variables. One commonly used method for this type of analysis is Multifactorial additive Analysis of Variance (ANOVA), which enables researchers to assess multiple dependent variables simultaneously. This approach is particularly useful when dealing with more than two independent groups, as it allows researchers to determine whether there is a statistically significant difference between the means of those groups.

The ANOVA is based on testing the null hypothesis that the mean values of all groups are equal. If the null hypothesis is rejected, it means that at least one pair of groups differs significantly on at least one variable, indicating that the independent variable explains a significant amount of variance in the dependent variable. However, ANOVA does not provide information about which group means are significantly different from each other. To uncover specific differences between three or more group means, post hoc tests are used. Post hoc tests, such as Duncan's Multiple Range test, measure specific differences between pairs of means.

Duncan's Multiple Range test is a post hoc test that compares all possible pairs of group means to identify which pairs differ significantly. This test is based on a ranking system that compares the distance between group means with the standard error of the mean. The test identifies significant differences between groups by dividing the mean square error by the number of degrees of freedom to calculate the standard error of the mean. This standard error is then compared to the difference between the means of each pair of groups. If the difference between the means is greater than the standard error, the two means are significantly different.

The statistical analysis of ANOVA and Duncan's Multiple Range test relies on several important assumptions: normality, homogeneity of variances, and independence. Normality refers to the requirement that the data follows a normal distribution, ensuring that the sample means are normally distributed. Homogeneity of variances implies that the variability among the groups being compared is approximately equal. Independence requires that each observation is independent to the others, meaning that the values of one observation do not influence or depend on the values of other observations.

It is important to note that obtaining reliable and valid results depends on adhering to these assumptions. When the assumption of normality is violated, it can lead to Type I or Type II errors. Type I error occurs when a significant difference is detected between groups when no true difference exists. Type II error occurs when a true difference between groups exists, but it is not detected as statistically significant. Violations of homogeneity of variances assumption can lead to biased estimates of group differences and impact the validity of the statistical tests. Therefore, assessing the assumptions of normality and homogeneity is crucial to ensure the reliability and validity of the results and minimize the potential for Type I and Type II errors.

Overall, statistical analysis, including ANOVA and post hoc analysis, is a crucial tool for making fair comparisons between groups and determining specific differences between means. They allow to identify significant relationships between variables and gain a deeper understanding of the mechanisms underlying their research questions.

In the present Doctoral thesis, we evaluated the influence of EMG-based visual biofeedback on users' performance [2, 3]. Prior conducting the analysis, normality and homogeneity of variances were verified. ANOVA was used for evaluating the effects of independent variables such as the type of test, test order, and individual on user performance. Next, we performed a Duncan's Multiple Range test to evaluate the differences in user performance with the presence or absence of two different types of feedback (EMG-based visual feedback and kinesthetic feedback from the motion exoskeleton) by calculating the difference between pairs of means.

References

1. Cisnal A, Perez-Turiel J, Fraile JC, Sierra D, De La Fuente E (2021) RobHand: a hand exoskeleton with real-time EMG-driven embedded control. Quantifying hand gesture recognition delays for bilateral rehabilitation. IEEE Access 9:137809–137823. https://doi.org/10.1109/ACCESS.2021.3118281
2. Cisnal A, Gordaliza P, Pérez Turiel J, Fraile JC (2023) Interaction with a hand rehabilitation exoskeleton in EMG-driven bilateral therapy: influence of visual biofeedback on the users' performance. Sensors 23:2048. https://doi.org/10.3390/s23042048

3. Cisnal A, Antolínez D, Turiel JP, Carlos Fraile J, De La Fuente E (2023a) A versatile embedded platform for implementation of biocooperative control in upper-limb neuromotor rehabilitation scenarios. IEEE Access 11:35726–35736. https://doi.org/10.1109/ACCESS.2023.3265898
4. Moreno-San Juan V, Cisnal A, Fraile J, Pérez-Turiel J, de la Fuente E (2021) Design and characterization of a lightweight underactuated RACA hand exoskeleton for neurorehabilitation. Rob Auton Syst 143:103828. https://doi.org/10.1016/j.robot.2021.103828
5. Alonso-Linaje G, Cisnal A, Fraile JC, Turiel JP (2024) Design and analysis of the M3Rob: a robotic platform for wrist and hand rehabilitation. IEEE Access 12:30472–30481. https://doi.org/10.1109/ACCESS.2024.3369908
6. Cisnal A, Martínez-Cagigal V, Alonso-Linaje G, Moreno-Calderón S, Turiel JP, Hornero R, Carlos J, Marinero F (2022a) An overview of M3Rob, a robotic platform for neuromotor and cognitive rehabilitation using augmented reality. In: XL Congreso Anual de La Sociedad Española de Ingeniería Biomédica, Valladolid, pp 180–183
7. McManus L, de Vito G, Lowery MM (2020) Analysis and biophysics of surface EMG for physiotherapists and kinesiologists: toward a common language with rehabilitation engineers. Front Neurol 11. https://doi.org/10.3389/fneur.2020.576729
8. Malmstrom E, Opton E, Lazarus R (1965) Heart rate measurement and the correlation of indices of arousal. Psychosom Med 27:546–556. https://doi.org/10.1097/00006842-196511000-00005
9. Meshkati N (1988) Heart rate variability and mental workload assessment. Adv Psychol 52:101–115. https://doi.org/10.1016/S0166-4115(08)62384-5
10. Novak D, Ziherl J, Olenšek A, Milavec M, Podobnik J, Mihelj M, Munih M (2010) Psychophysiological responses to robotic rehabilitation tasks in stroke. IEEE Trans Neural Syst Rehabil Eng 18:351–361. https://doi.org/10.1109/TNSRE.2010.2047656
11. Ekman P, Levenson RW, Friesen WV (1983) Autonomic nervous system activity distinguishes among emotions. Science 1979(221):1208–1210. https://doi.org/10.1126/SCIENCE.6612338
12. Halaki M, Ginn K (2012) Normalization of EMG signals: to normalize or not to normalize and what to normalize? In: Naik GR (ed) Computational intelligence in electromyography analysis—a perspective on current applications and future challenges. InTech, Rijeka, Croatia, pp 175–194. https://doi.org/10.5772/3315
13. Chowdhury RH, Reaz MBI, Bin Mohd Ali MA, Bakar AAA, Chellappan K, Chang TG (2013) Surface electromyography signal processing and classification techniques. Sensors 13:12431–12466. https://doi.org/10.3390/s130912431
14. Conforto S (2009) The role of the sEMG signal processing in the field of the human movement analysis. In: Dösel O, Schlegel WC (eds) World congress on medical physics and biomedical engineering, 7–12 Sept 2009, Munich, Germany. IFMBE Proceedings, vol 25/9. Springer, Berlin, Heidelberg, pp. 523–526. https://doi.org/10.1007/978-3-642-03889-1_140
15. Luca CJ De (2002) Surface electromyography: detection and recording. Delsys Incorporated
16. Lehman GJ, McGill SM (1999) The importance of normalization in the interpretation of surface electromyography: a proof of principle. J Manipulative Physiol Ther 22:444–446. https://doi.org/10.1016/S0161-4754(99)70032-1
17. Lucas L, DiCicco M, Matsuoka Y (2016) An EMG-controlled hand exoskeleton for natural pinching. J Rob Mech 16:482–488. https://doi.org/10.20965/jrm.2004.p0482
18. Serpelloni M, Tiboni M, Lancini M, Pasinetti S, Vertuan A, Gobbo M (2016) Preliminary study of a robotic rehabilitation system driven by EMG for hand mirroring. In: 2016 IEEE international symposium on medical measurements and applications (MeMeA). IEEE, Benevento, Italy, pp 1–6. https://doi.org/10.1109/MeMeA.2016.7533730
19. Polygerinos P, Galloway KC, Sanan S, Herman M, Walsh CJ (2015a) EMG controlled soft robotic glove for assistance during activities of daily living. In: 2015 IEEE international conference on rehabilitation robotics (ICORR). IEEE, Singapore, pp 55–60. https://doi.org/10.1109/ICORR.2015.7281175
20. Chen M, Ho SK, Zhou HF, Pang PMK, Hu XL, Ng DTW, Tong KY (2009) Interactive rehabilitation robot for hand function training. In: 2009 IEEE international conference on rehabilitation robotics (ICORR), Kyoto, Japan, pp 777–780. https://doi.org/10.1109/ICORR.2009.5209564

21. Ho NSK, Tong KY, Hu XL, Fung KL, Wei XJ, Rong W, Susanto E (2011) An EMG-driven exoskeleton hand robotic training device on chronic stroke subjects. IEEE Int Conf Rehabil Robot 2011:149–153. https://doi.org/10.1109/ICORR.2011.5975340
22. Scotto Di Luzio F, Simonetti D, Cordella F, Miccinilli S, Sterzi S, Draicchio F, Zollo L (2018) Bio-cooperative approach for the human-in-the-loop control of an end-effector rehabilitation robot. Front Neurorobot 12:1–12. https://doi.org/10.3389/fnbot.2018.00067
23. Novak D, Mihelj M, Ziherl J, Olenšek A, Munih M (2011) Psychophysiological measurements in a biocooperative feedback loop for upper extremity rehabilitation. IEEE Trans Neural Syst Rehabil Eng 19:400–410. https://doi.org/10.1109/TNSRE.2011.2160357
24. Mihelj M, Novak D, Munih M (2009) Emotion-aware system for upper extremity rehabilitation. In: 2009 Virtual rehabilitation international conference, VR 2009 160–165. https://doi.org/10.1109/ICVR.2009.5174225
25. Guerrero CR, Fraile Marinero JC, Turiel JP, Muñoz V (2013) Using "human state aware" robots to enhance physical human-robot interaction in a cooperative scenario. Comput Methods Programs Biomed 112:250–259. https://doi.org/10.1016/j.cmpb.2013.02.003
26. Mandryk RL, Atkins MS (2007) A fuzzy physiological approach for continuously modeling emotion during interaction with play technologies. Int J Hum Comput Stud 65:329–347. https://doi.org/10.1016/j.ijhcs.2006.11.011
27. Gümüslü E, Erol Barkana D, Köse H (2020) Emotion recognition using EEG and physiological data for robot-assisted rehabilitation systems. In: Gümüslü E, Barkana DE, Köse H (eds) Companion publication of the 2020 international conference on multimodal interaction. Association for computing machinery, New York, pp 379–387. https://doi.org/10.1145/3395035.3425199
28. Khezri M, Firoozabadi M, Sharafat AR (2015) Reliable emotion recognition system based on dynamic adaptive fusion of forehead biopotentials and physiological signals. Comput Methods Programs Biomed 122:149–164. https://doi.org/10.1016/j.cmpb.2015.07.006
29. Picard RW, Vyzas E, Healey J (2001) Toward machine emotional intelligence: analysis of affective physiological state. IEEE Trans Pattern Anal Mach Intell 23:1175–1191. https://doi.org/10.1109/34.954607
30. Hariharan A, Adam MTP (2015) Blended emotion detection for decision support. IEEE Trans Hum Mach Syst 45:510–517. https://doi.org/10.1109/THMS.2015.2418231
31. Kim J, André E (2008) Emotion recognition based on physiological changes in music listening. IEEE Trans Pattern Anal Mach Intell 30:2067–2083. https://doi.org/10.1109/TPAMI.2008.26
32. Koelstra S, Mühl C, Soleymani M, Lee JS, Yazdani A, Ebrahimi T, Pun T, Nijholt A, Patras I (2012) DEAP: A database for emotion analysis; using physiological signals. IEEE Trans Affect Comput 3:18–31. https://doi.org/10.1109/T-AFFC.2011.15
33. Liu C, Conn K, Sarkar N, Stone W (2008) Physiology-based affect recognition for computer-assisted intervention of children with autism spectrum disorder. Int J Hum Comput Stud 66:662–677. https://doi.org/10.1016/j.ijhsc.2008.04.003
34. Badesa FJ, Morales R, Garcia-Aracil N, Sabater JM, Casals A, Zollo L (2014) Auto-adaptive robot-aided therapy using machine learning techniques. Comput Methods Programs Biomed 116:123–130. https://doi.org/10.1016/j.cmpb.2013.09.011
35. Yang E, Dorneich MC (2015) The effect of time delay on emotion, arousal, and satisfaction in human-robot interaction. In: Proceedings of the human factors and ergonomics society 59th annual meeting 2015, pp 443–447. https://doi.org/10.1177/1541931215591094
36. Tharwat A (2018) Classification assessment methods. Appl Comput Inform 17:168–192. https://doi.org/10.1016/j.aci.2018.08.003
37. Hossin M, Sulaiman MN (2015) A review on evaluation metrics for data classification evaluations. Int J Data Min Knowl Manag Process 5:1–11. https://doi.org/10.5121/ijdkp.2015.5201
38. Grandini M, Bagli E, Visani G (2020) Metrics for multi-class classification: an overview. ArXiv 1–17. https://doi.org/10.48550/arXiv.2008.05756
39. Li G, Member S, Schultz AE, Kuiken TA, Member S (2010) Quantifying pattern recognition—based myoelectric control of multifunctional transradial prostheses. IEEE Trans Neural Syst Rehabil Eng 18:185–192. https://doi.org/10.1109/TNSRE.2009.2039619

40. Ding H, Trajcevski G, Scheuermann P, Wang X, Keogh E (2008) Querying and mining of time series data: experimental comparison of representations and distance measures. Proc VLDB Endowment 1:1542–1552. https://doi.org/10.14778/1454159.1454226

Chapter 4
Results

In this chapter, the most relevant results of the Doctoral Thesis are summarized. Section 4.1 introduces the designed embedded solutions for physiological signal acquisition, while Sect. 4.2 elaborates on the development of biocooperative control strategies. The results of the performance assessment are presented in Sect. 4.3.

4.1 Embedded Systems for Physiological Signal Acquisition

This section presents two embedded systems for physiological signal acquisition that have been developed. The first one was specially designed for the implementation of rehabilitation therapies based on sEMG signals [1]. The second solution features a wearable design with enhanced versatility, enabling the acquisition of a broader range of physiological signals to facilitate the development of multimodal biocooperative controls [2].

The primary objective behind the design and development of embedded electronic systems for physiological signal acquisition is to overcome certain limitations identified in prior studies. Specifically, these include the high cost of such systems, which is a major obstacle to universal access, as well as the use of computers as the data processing platforms, which presumably increases the latency time of the entire rehabilitation platform. As a solution to these challenges, custom-made systems have been developed with the goal of minimizing overall costs. Additionally, these systems incorporate a real-time microcontroller (MCU) for advanced on-board processing.

Both systems are based on the TMS320F28069M MCU (Texas Instruments, Texas, USA), whose Harvard architecture is optimized to perform real-time tasks. The MCU is characterized by a two-core architecture resulting in a large system bandwidth. C28x core and CLA (Control Law Accelerator) execute code independently and interface using a specialized data bus (CLA Bus). The high efficiency 32-bit C28x core runs up to 90 kHz. It is also equipped with 256 Kb Flash embedded

A. Cisnal de la Rica, *Development of Biocooperative Control Strategies for Neuromotor Rehabilitation Robotic Platforms*, Springer Theses,
https://doi.org/10.1007/978-3-032-02485-5_4

memory, 100 KB of RAM and 2 KB of one-time programmable (OTP) ROM. Additionally, the MCU integrates multiple peripherals such as a 12-bit analog-to-digital converter (ADC) with a sampling rate up to 3.46 MSPS (mega samples per second), serial port communication peripherals (SPI, i2C, UART …), ePWM (enhanced Pulse Width Modulation) modules and timers. Normal current consumption is 245 mA, allowing low-power operating modes.

4.1.1 EMG Acquisition System

EMG-driven control strategies are widely used in rehabilitation robotics. Muscular electrical activity resulting from MUAP superposition (raw sEMG signals) can be acquired by surface electrodes and proper conditioning circuits. Therefore, a custom-made application-specific integrated circuit (ASIC) was designed to capture and preprocess the sEMG signals, which are subsequently converted by an ADC for its transmission to the TMS320F28069M MCU, especially dedicated to real-time processing operations.

The 2-channel EMG data acquisition ASIC developed is characterized by a 24-bit resolution differential channels and 112 DB of dynamic range (DR). The channels consist of an instrumentation amplifier of gain 50 followed by an RC low-pass filter with a cutoff frequency of 150 Hz. The channels are designed to compensate differential input offset to prevent the instrumentation amplifier from saturating. Additionally, the ASIC has the MCP3912 (Microchip Technology Inc., AZ, USA) Analog Front End (AFE), which is characterized by synchronous Delta-Sigma ADCs, which interfaces with the TMS320F28069M MCU using SPI communication and other digital control signals.

The 4-layer Printed Circuit Board (PCB) has split ground planes to separate analog, digital, and power circuitries to ensure signal integrity (Fig. 4.1a). Discrete Surface Mounted Device (SMD) components are placed on the top and bottom layers, resulting in a board size is 50.8 $\times$ 33 mm, with an active area of 10 cm^2 (Fig. 4.1b). The PCB was purposely designed to enable direct fit of the LAUNCHXL-F28069M MCU, thereby eliminating the need of wiring (Fig. 4.1c). The chip consumes 3 mW from 3.3 V DC power supply.

The main characteristics of the proposed EMG signal acquisition system are outlined in Table 4.1. The system is powered by a 5 V supply derived from the MCU, despite the admissible power supply ranging from 3 to 12 V. Additionally, the Delta-Sigma ADCs are fully configurable, and he configuration details specifically established in this thesis are presented in Table 4.1.

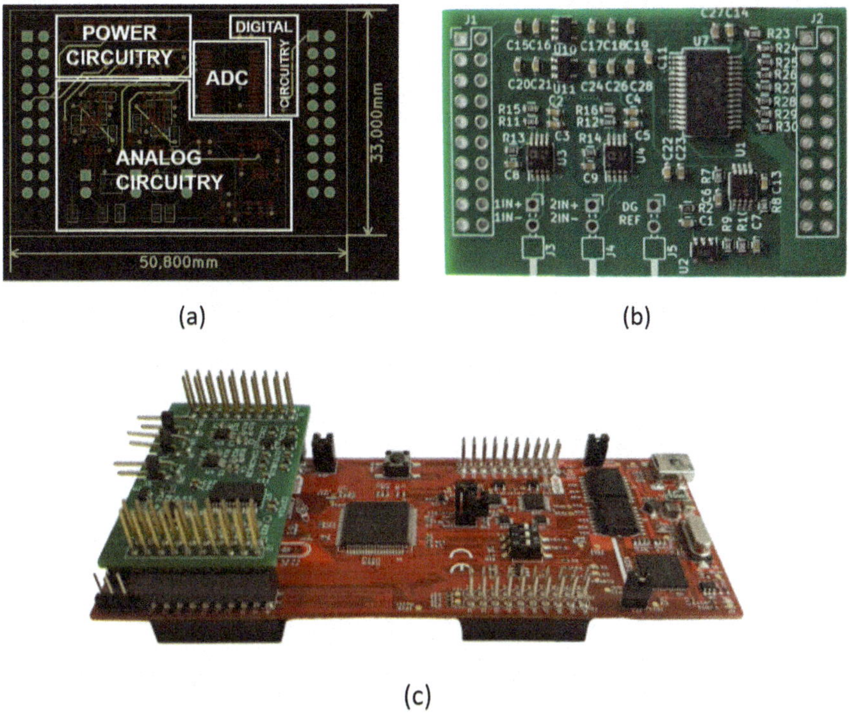

Fig. 4.1 EMG acquisition system layout **a** PCB layout **b** final solution **c** board plugged-in the MCU

4.1.2 Multimodal Acquisition System

The proposed embedded wearable system integrates multiple sensors modalities, high-efficiency real-time MCUs, and wireless communication, providing a highly flexible and capable platform for the development of biocooperative controllers. The system is battery-powered and integrates five sensors (IMU, ECG, EMG, GSR, and SKT), which directly interface with the TMS320F28069M MCU for real-time data processing and control (Fig. 4.2). The recorded and processed data can be transmitted by a CC2650 (Texas Instruments, TX, USA) Bluetooth low energy (BLE) MCU. Additionally, a JTAG port allows temporary access to the MCUs.

An ICM-20948 9-axis MEMS device (InvenSense, CA, USA) allows to track the motion of the user wearing the device. It consists of a gyroscope, an accelerometer and a compass with programmable sensitivities and filters. It also embeds a digital motion processor (DMP) aimed to reduce the computational load associated to the motion algorithm from the MCU, thus improving system performance. The ICM-20948, which is powered at 1.8 V, communicates with the TMS320F28069M using Fast Mode i2C.

Table 4.1 Technical specifications of the EMG acquisition system, encompassing general characteristics and configuration parameters of the delta-sigma ADCs

General characteristics	Values
Supply voltage	3.3–12 V/5 V[a]
Supply voltage	5 V
Total current consumption	0.9 mA
Active area	10 cm^2
Board dimensions	50.8 × 33 mm
EDS input protection	17 kV
Acquisition channels	2 ch
ΔΣ-ADC Performance	**Values**
Frequency sample	200 Hz[a]
Acquisition bandwidth	195 Hz[a]
Digital signal resolution	24 bits[a]
Operating frequency	3.3 MHz[a]
Differential input impedance	232 KΩ[a]
DD CMRR	100 dB
Signal-to-noise ratio (SNR)	94 dB

[a]Established configuration

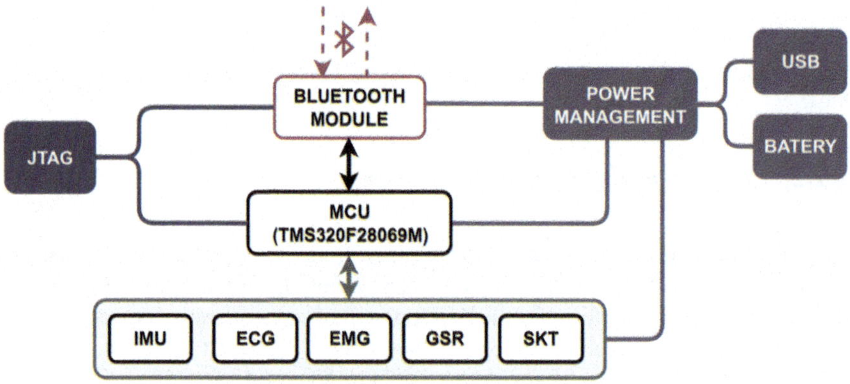

Fig. 4.2 High level block diagram of the proposed multimodal embedded system [2]

The fully integrated single-lead ECG front-end AD8232 (Analog Devices Inc., MA, USA), powered at 3.3 V, amplifies, and filters the ECG signals. It also incorporates an AC lead-off detection mode. The recorded ECG signals are analog-to-digital converted in the MCU with 12-bit resolution and two additional digital signals interface with the MCU to indicate whether the snap-lead pre-gelled electrodes are properly attached and connected using a jack connector.

A 2-channel electromyographic signal acquisition module is also integrated into the printed circuit board. The electronic circuit of the module is the same as described in Sect. 4.1.1. The GSR measurements are based on a low constant voltage technique. A low electrical potential is applied between two electrodes attached to two hand fingers and the resulting current flow is amplified by the LM324 quadruple operational amplifier (Texas Instruments, TX, USA) and then, filtered. The analog output is transmitted to the MCU to be ADC.

The MLX90614 infrared (IR) thermometer (Melexis, Belgium) allows to measure the body temperature without physical contact and thus, facilitating its continuous monitoring. The MLX90614 is characterized by an IR-sensitive thermopile detector chip and conditioning circuit, which includes a low-noise amplifier, a 17-bit ADC and a digital signal processing unit. It transmits the temperature measurements with up to 0.02 °C resolution to the MCU via i2C and it is powered at 3.3 V.

The CC2650 MCU is a system-on-chip (SOC) that provides an ultralow power BLE using a 2.4 GHz RF transceiver. The MCU is built on an ARM® Cortex®-M3 processor for the application layer and BLE protocol stack management and on ARM Cortex®-M0 processor for the autonomous low-level radio control and processing related to the physical and link layers. The CC2650 MCU interfaces with the TMS320F28069M using one of the following serial communications: i2C, SPI or UART. No external antenna is required since the PCB integrates one on the top cooper layer, flowing TI's design specifications.

The device is powered by a battery or via a USB-C port if the battery is being recharged. The system integrates a MCP73831 dedicated integrated circuit (Microchip Technology Inc., AZ, USA), which is based on a constant voltage/constant current (CVVC) charging method. The charge management controller continuously monitors the battery voltage and starts its recharge if the voltage drops below a threshold. A 3.7 V 3500 mAh rechargeable lithium-ion polymer battery (68 $\times$ 55 $\times$ 7 mm) provides an estimated life of 5 h, considering that a maximum power consumption of 1250 mW.

The 2-layer PCB was designed to optimize signal integrity, and its layout is based on the different submodules (Fig. 4.3a). The power planes are split, and 12 mil wide traces are used, except for 24 mil wide power signals. For reducing assembling costs, all components are placed on the top layer, except for the thermometer. The resulting board size is 63 $\times$ 83 mm, including three holes for screw fixing to the box (Fig. 4.3b).

4.2 Biocooperative Rehabilitation Systems

This section aims to present three biocooperative control strategies for upper-limb rehabilitation using the acquisition systems described in the previous Sect. 4.1. An EMG-driven bilateral control for hand robotic rehabilitation is detailed in Sect. 4.2.1. In Sect. 4.2.2, a VR-based exergame using hand gesture detection and arm tracking is presented. Lastly, an AAN control for robotic wrist rehabilitation therapy is described in Sect. 4.2.3.

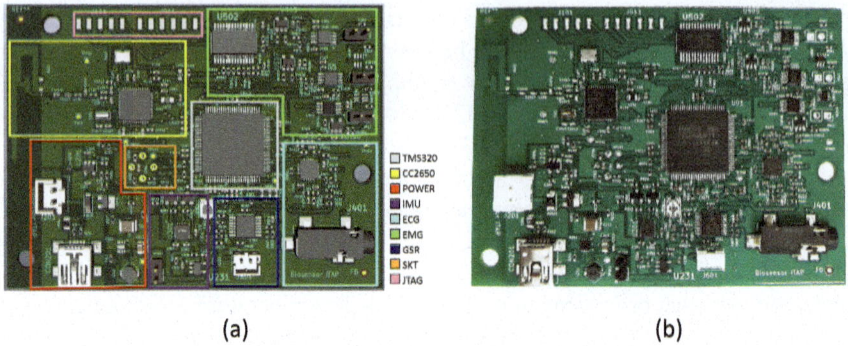

Fig. 4.3 Multimodal acquisition system **a** PCB layout **b** final solution [2]

4.2.1 EMG-Driven Control for Hand Robotic Rehabilitation

A threshold EMG-driven control has been proposed for conducting bilateral training with the RobHand rehabilitation robot (Sect. 3.1.1). The exoskeleton is worn on the impaired hand to provide assistance in motion, while surface electrodes are attached to the forearm muscles and olecranon of the healthy limb for recording the EMG signals (Fig. 4.4).

The embedded system described in Sect. 4.1.1 was used to record the sEMG signals of the ED and the FDS muscles, responsible for the extension and flexion of the hand fingers. The acquired signals were processed by the MCU to recognize the gesture performed by the healthy hand (rest, open or close), generating the proper pulse width modulation (PWM) signals so the rehabilitation exoskeleton placed on the paretic hand moves to replicate the detected gesture (Fig. 4.5).

The EMG signals, which are recorded at 200 Hz, are filtered for baseline noise and electromagnetic interferences removal and then, rectified and normalized to compare the electrical activity between muscles (Fig. 4.6). A notch center frequency (50 Hz cuff-off frequency; 20 Q-factor) and a high-pass FIR filter (0.01 Hz stopband

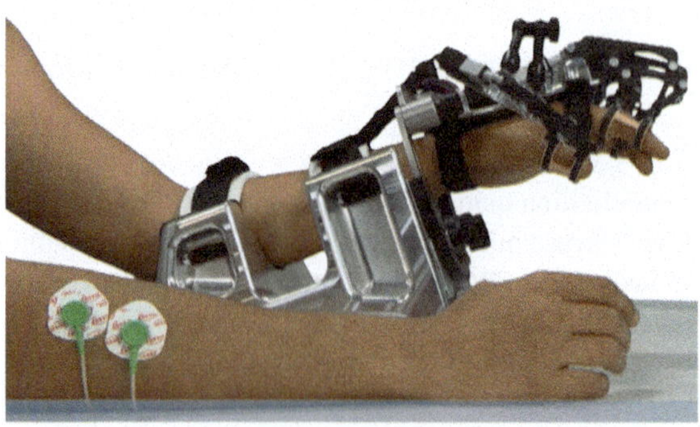

Fig. 4.4 User performing EMG-driven bilateral therapy using the RobHand rehabilitation platform. The exoskeleton is worn on the impaired hand, while disposable electrodes are attached to the forearm muscles of the healthy hand to capture EMG signals [3]

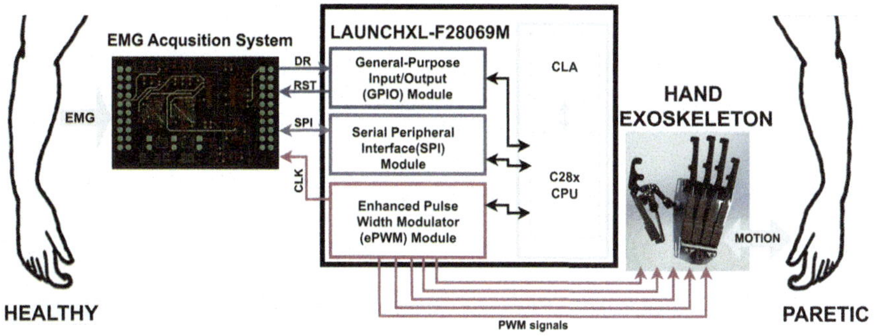

Fig. 4.5 Simplified block diagram of the implementation of the EMG-driven control on the RobHand platform for assisted bilateral therapy [1]

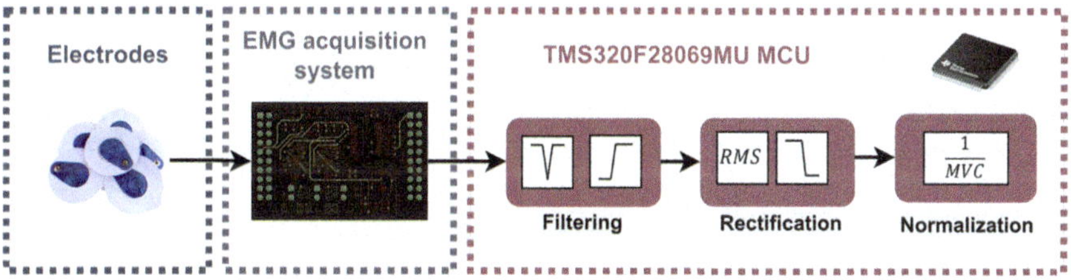

Fig. 4.6 sEMG signal recording and processing, including filtering, rectification, and normalization [1]

frequency; 10 Hz passband frequency: 80 dB minimum stopband attenuation; 0.1 dB maximum passband ripple) are applied to the raw EMG signals. The rectification is based on a 10-point RMS and a low-pass FIR filter (1 kHz passband edge frequency; 2 Hz stopband edge frequency; 4 dB maximum passband ripple; 10 dB minimum stopband attenuation), resulting in a down-sampling of 20 Hz. The normalization is performed with respect to the MVC [4].

The MVC values and EMG-based thresholds are computed in a calibration, which is user and rehabilitation session-specific. In the calibration procedure, subjects are asked to relax, open, and close their hand for 8 s each (Fig. 4.7). The MVC values (MVC_{ED} and MVC_{FDS}) are calculated as the maximum value of their corresponding rectified EMG signal (rEMG), as indicated in Eq. (4.1).

$$MVC_{FDS/ED} = \max\left(rEMG_{FDS/ED}\right) \tag{4.1}$$

The extensor and flexor threshold, defined as ε and μ, correspond to the maximum limit values corresponding to muscular deactivation of the FDS and ED muscles, respectively. They are calculated following Eqs. (4.2) and (4.3).

$$\varepsilon = \frac{\min(rEMG_{ED})}{MVC_{ED}} + 0.1 \tag{4.2}$$

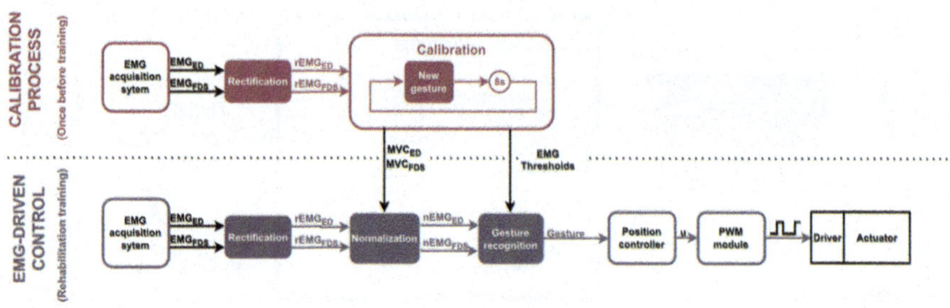

Fig. 4.7 Control loop implemented for the calibration process and threshold-based EMG-driven control [1]

$$\mu = \frac{\min(rEMG_{FDS})}{MVC_{FDS}} + 0.1 \tag{4.3}$$

The gesture recognition module (Fig. 4.7) only depends on the instantaneous normalized EMG signals (nEMG) and the two EMG-based thresholds (μ and ε). Considering that A, B and C are defined according to Eqs. (4.4–4.6).

$$A \rightarrow nEMG_{ED} > \varepsilon \tag{4.4}$$

$$B \rightarrow nEMG_{FDS} > \mu \tag{4.5}$$

$$C \rightarrow nEMG_{ED} > nEMG_{FDS} \tag{4.6}$$

The gesture (rest, open or close) is updated every 50 ms based on Eqs. (4.7–4.9) Note that for any combination of inputs, only one output (REST, OPEN or CLOSE) is true.

$$REST = \overline{A} \cdot \overline{B} \tag{4.7}$$

$$OPEN = A \cdot (\overline{B} + C) \tag{4.8}$$

$$CLOSE = B \cdot (\overline{A} + \overline{C}) \tag{4.9}$$

The rest gesture is recognized when the normalized signals are lower than their corresponding EMG-based threshold. The open gesture is detected when the normalized sEMG signal from the ED muscle is larger than both the extensor threshold (ε) and the normalized signal from the FDS muscle if it exceeds the flexor threshold (μ). Analogously, the closed gesture is detected when the normalized sEMG signal from the FDS muscle is larger than the flexor threshold (μ) and the normalized signal from ED muscle in case it exceeds the extensor threshold (ε). Figure 4.8 shows the signal

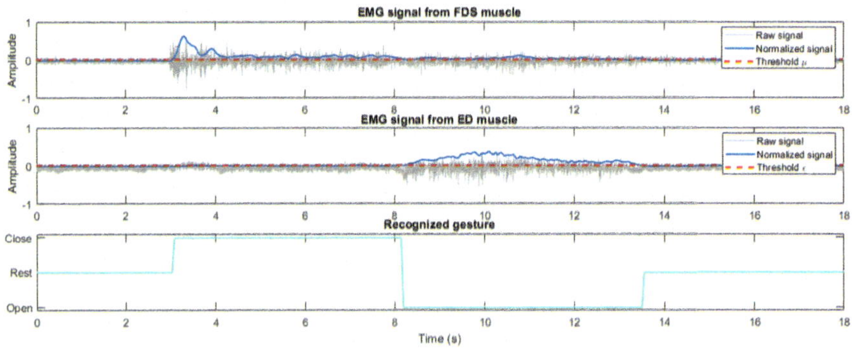

Fig. 4.8 Threshold EMG-driven control: raw EMG signals of ED and FDS muscles, normalized signals, EMG-based threshold, and recognized gestures [3]

processing of the presented EMG-driven control: raw and normalized signals based on user's residual muscle activity, thresholds, and recognized gestures.

Incorporation of EMG-based visual feedback to the presented EMG-driven bilateral therapy is feasible. The proposed biofeedback approach involves two variable-length bars, labelled as 'Opening force' and 'Closing force'. These bars are named so as to correspond to the extensor and flexor muscle activity and have been designed to enhance the user-friendliness of the therapy (Fig. 4.9).

The length of the bar indicates the instantaneous value of the normalized sEMG signals ($nEMG_{ED}$ and $nEMG_{FDS}$) and their color indicates the recognized gesture (Table 4.2) according to Eqs. (4.7–4.9). The visual feedback is updated at 20 Hz, corresponding to the frequency of the normalized EMG signals and gesture recognition rate.

Fig. 4.9 EMG-based visual feedback consisting of two variable length bars [3]

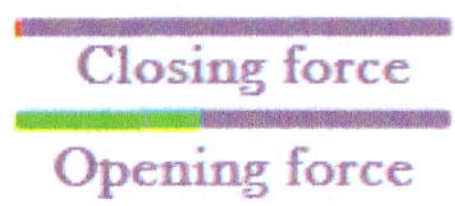

Table 4.2 Color of the bars of the EMG-based visual feedback depending on the recognized gesture

Recognized gesture	Opening force bar	Closing force bar
Rest	Red	Red
Closed	Red	Green
Opened	Green	Red

4.2.2 EMG&IMU-Based Control for Upper-Limb Rehabilitation

A VR-based exergame for upper-limb rehabilitation was developed using the multimodal embedded system (Sect. 4.1.2). The system is worn on the user's arm to track the arm and hand movements, and thus enabling interaction with the virtual environment. Two pairs of surface electrodes are attached to the ED and FDS forearm muscles, along with one reference electrode to the olecranon, for recording EMG signals.

The objective of the exergame is to collect as many coins as possible within a fixed time by moving the arm towards the coins with a relaxed or open hand, and then closing the hand to take the coin (Fig. 4.10). The difficulty of the game is adjusted online based on the elapsed time from each coin collection. Modifications are made to the degree of hand closure as well as size and relative position of the new coin respect to the last one to maintain patient motivation.

The sEMG signals captured from the ED and FDS muscles as well as the linear acceleration and angular velocity data from the ICM-20948 are transmitted to the TMS320F228069M MCU. Real-time algorithms for arm orientation estimation and hand gesture detection (open, rest and close) are executed on the MCU. The gesture and orientation are then transmitted via BLE at a rate of 20 Hz to a PC, which updates the limb movements in the VR environment, allowing the hand to interact with the virtual coins (as shown in Fig. 4.11). The VR scenario was developed using the Unity game engine.

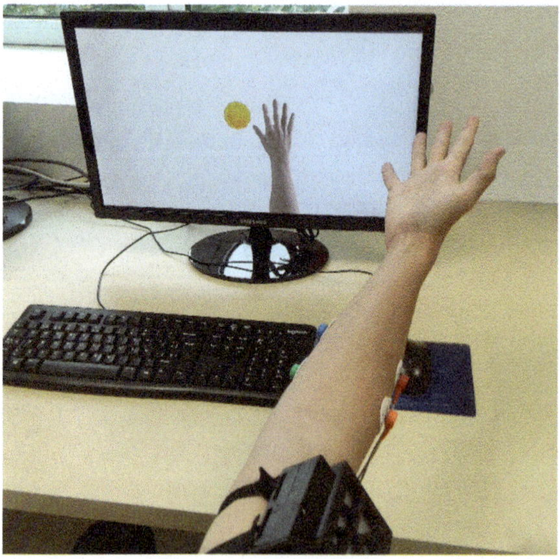

Fig. 4.10 Upper-limb rehabilitation using a VR-based exergame. The embedded platform is placed on the arm user and two pairs of surface electrodes are attached to the ED and FDS muscles and the reference electrode is attached to the olecranon [2]

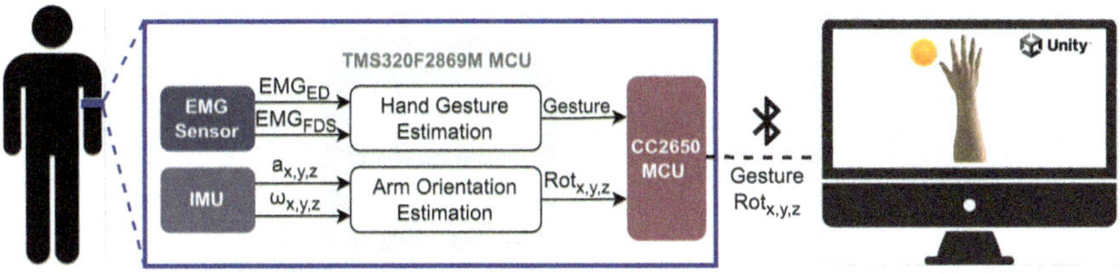

Fig. 4.11 Schematic diagram of the VR-based exergame for upper-limb rehabilitation [2]

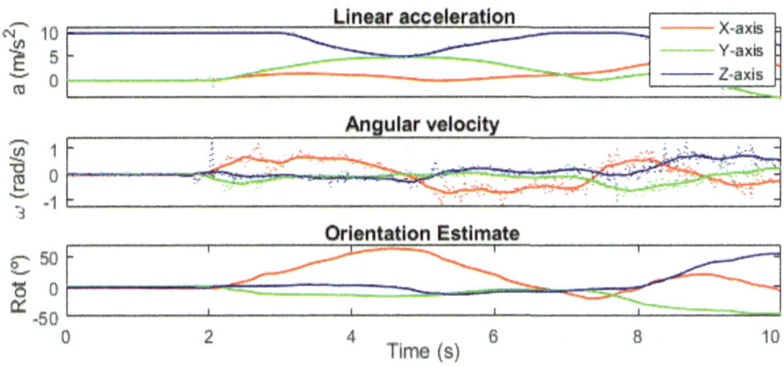

Fig. 4.12 Recorded linear acceleration (top plot), recorded angular velocity (middle plot) and estimated orientation (bottom plot). Solid lines indicate raw data and dashed lines indicate filtered data [2]

The hand gesture recognition algorithm is the one presented in Sect. 4.2.1. Regarding arm orientation estimation, the 16-bits 3 axis gyroscope and accelerometer of the ICM-20948 are set to full-scale range of ± 250 dps (degrees per second) and ± 2 g (19.6 m/s^2), resulting in 131 LSB/dps and 16,384 LSB/g ADC resolution, respectively. The sample rate is configured to 100 Hz and 400 kHz Fast Mode i2C is selected to transmit both linear acceleration and angular velocity to the MCU. On the MCU, the orientation estimation is carried out by implementing the model described in [5]. The accelerometer and gyroscope readings are entered to this model, which is based on an indirect Kalmar filter. The linear acceleration, angular velocity and estimated arm orientation readings are shown in Fig. 4.12.

4.2.3 Assist-as-Needed Control for Wrist Robotic Rehabilitation

An AAN control strategy that adapts the level of assistance of a wrist rehabilitation robot (Sect. 3.1.2) based on the emotional state and the motor performance of the patient has been proposed. Biomechanical measurements from the encoder and force

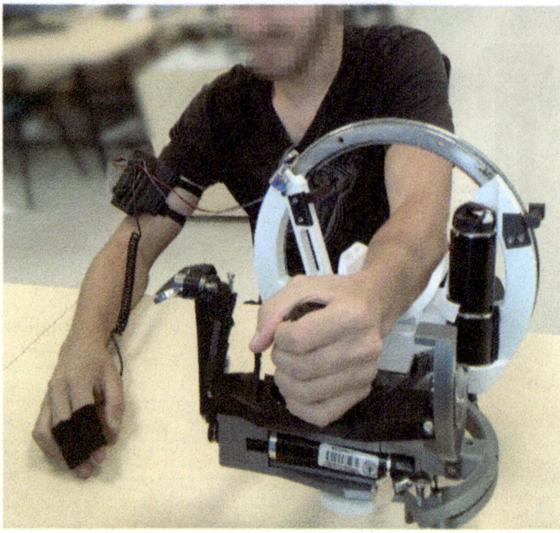

Fig. 4.13 User undergoing rehabilitation using the wrist robotic platform based on an AAN control strategy, while wearing the embedded platform on the arm to register SKT. To record ECG signals, disposable electrodes are attached to the user's torso, and electrodes are placed on the fingertips of the hand to capture GSR [2]

sensor embedded on the rehabilitation robot are used to determine the motor performance. ECG, GSR and SKT values recorded with the embedded system (Sect. 4.1.2) are used to estimate the emotional state of the user considering the two-dimensional model (arousal and valence).

The user must wear the embedded system to register their SKT with the IR thermometer. Two disposable pre-gelled electrodes are attached to the torso, and one on the umbilical region of the user for ECG recording. GSR measurement is accomplished by placing two electrodes on the index and middle fingertips (Fig. 4.13).

The AAN paradigm is applied to a closed-loop admittance control with a reference trajectory. The mechanical admittance (Y) of a system is defined as the ratio of its displacement (x) and its force (F), as expressed in Eq. (4.10). This definition is usually related to a mass-spring-damper system, characterized by a mass (m), a stiffness (k) and a viscous damping (b) [6].

$$Y(s) = \frac{X}{F} = \frac{1}{ms^2 + bs + \mathrm{k}} \tag{4.10}$$

The objective of an admittance control is to shape the mechanical admittance of a device such that it possesses desired characteristics. Hence, M, B, and K parameters are updated (increase, decrease, or no change) every second according to assistance level of the robot. The level of assistance is determined by a two-stage fuzzy logic model (Fig. 4.14), which is similar in nature to that presented by Mihelj et al. [7].

Before the training session, the user must relax for 30 s to determine their physiological resting values. The first stage of the fuzzy logic evaluates the motor performance based on the biomechanical information (force and motion) provided by the force sensor and encoders of the rehabilitation robot. The first stage also determines

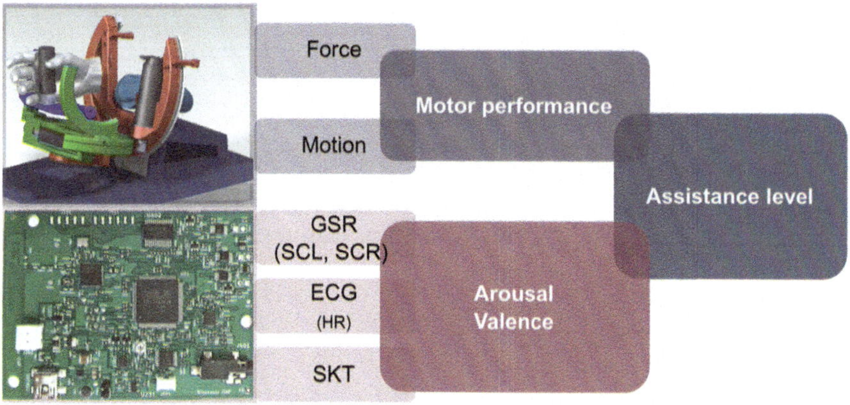

Fig. 4.14 AAN control strategy based on a two-stage fuzzy logic model which considers the motor performance and emotional state of the user [2]

whether to increase, decrease or maintain the arousal and valence in function of the variations in HR, SCL, SCR frequency and SKT, all of them normalized with respect to their resting values. The algorithm that determines the changes in valence and arousal is based on the ones presented by Mandryk et al. [8] and Guerrero et al. [9].

The Pan–Tompkins's algorithm [10] was used to detect the R events, and hence the heart rate from the ECG signals recorded at 500 Hz (Fig. 4.15). The algorithm applies a series of filters to the ECG signal and squares the filtered signal to amplify the QRS contribution and then, uses adaptive thresholds to detect the peaks of the processed ECG signals.

The skin conductance (SC), recorded by the GSR module at a 225 Hz, is low pass filtered (10 Hz cut-off frequency) to eliminate the high-frequency noise, which may cause false-positive detection of phasic events. The tonic component (SCL) and phasic component (SCR) are extracted from the filtered SC (Fig. 4.16) using a deconvolution technique [11].

In the second stage, the decision involves determining whether to increase, maintain or decrease the assistance level based on the variations in arousal, valence, and motor performance identified in the previous stage. The physiological signals (GSR, ECG and SKT) are processed in the MCU and the extracted features (HR, SCL, SCR, and SKT) are sent to the PC via BLE. These features are then combined with the

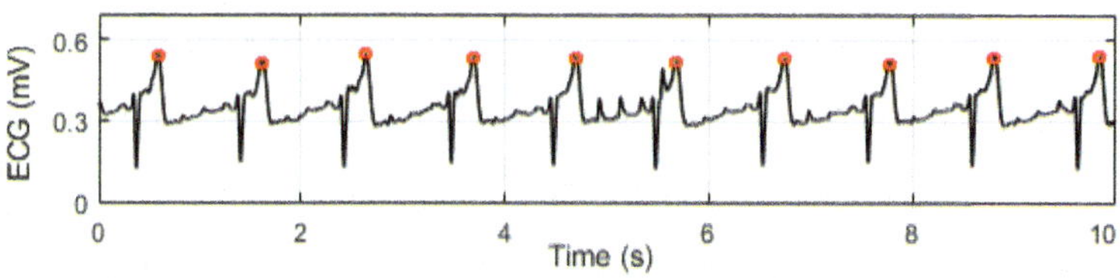

Fig. 4.15 R events of an ECG signal from a person at rest [2]

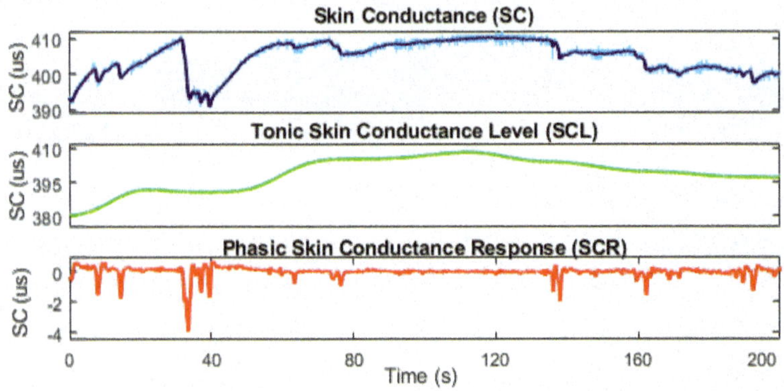

Fig. 4.16 Raw and filtered skin conductance and its tonic (SCL) and phasic (SCR) components [2]

biomechanical data (force, position, and velocity) are entered into the fuzzy model. This model estimates the assistance level and updates the mechanical admittance of the robot. The force, position and velocity data are sent from the robot to the PC, while the admittance parameters, namely K, B and M, are transmitted from the PC to the robot (Fig. 4.17).

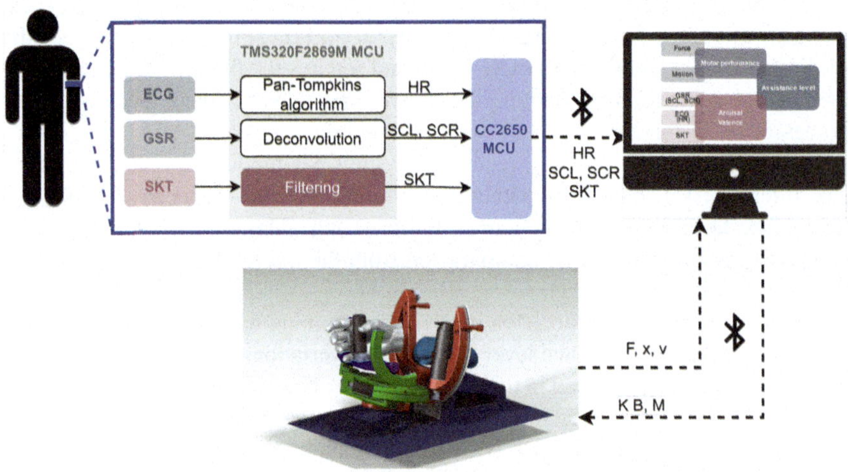

Fig. 4.17 Schematic diagram of the AAN control strategy for wrist rehabilitation

4.3 Performance Assessment

The evaluation of the HRI of the EMG-driven hand exoskeleton was performed by assessing system responsiveness and gesture recognition accuracy (Sect. 3.4.2). The experimental setup, as detailed in Sect. 3.3.1, was used. The results of the time delay analysis are presented in Fig. 4.18. The mean and standard deviation of the MST and the MOT were 0.48 ± 0.59 s and 0.55 ± 0.60 s, respectively. The MCT was 1.90 ± 1.65 s, varying from 0.98 s (from close to rest gesture) to 3.42 s (from open to close gesture). The MCT is dependent on the actuators' speed and was deemed to be sufficiently long to ensure user safety during the rehabilitation. The motion-completion rate was 100%, as MCT did not exceed the predefined time limit of 5 s in any case.

The confusion matrix that presents the performance of the EMG-based gesture recognition control is depicted in Fig. 4.19. The confusion matrix was obtained by comparing the actual hand gestures recorded by the 5DT Data Glove and the corresponding gestures recognized by the EMG-based control. Before calculating the confusion matrix, data was cleaned to eliminate the gesture transition period (time needed to move the hand from one gesture to another), to avoid false negatives. The gesture transition period was considered 0.9 s (maximum MST value) starting when a transient period is detected by the glove. The overall accuracy of the gesture recognition is 97% and misclassifications during the gestures are minimal (fewer than 0.06 s). This means that in 98.4% of cases, the error is not noticeable in the exoskeleton, as the actuators do not start moving in the opposite direction to the intended movement.

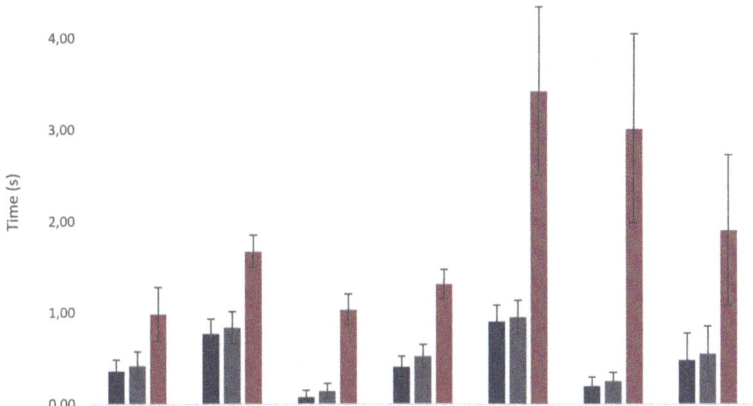

	CloseToRest	RestToClose	RestToOpen	OpenToRest	OpenToClose	CloseToOpen	Mean
Motion-selection time	0,36	0,77	0,08	0,41	0,90	0,20	0,48
Motion-onset time	0,42	0,84	0,15	0,52	0,95	0,25	0,55
Motion-completion time	0,98	1,67	1,03	1,31	3,42	3,01	1,90

Fig. 4.18 Time delay analysis of the EMG-driven bilateral control with the hand exoskeleton [1]

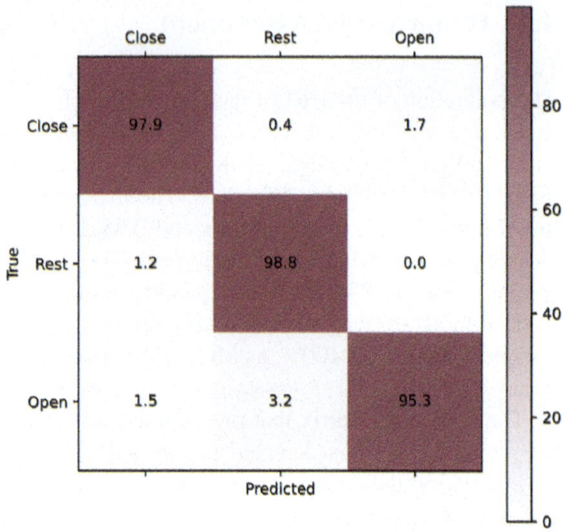

Fig. 4.19 Confusion matrix of the EMG-based gesture recognition [1]

The influence of the EMG-based visual feedback on the user performance when performing EMG-driven bilateral assisted therapies was also evaluated. The experimental setup was detailed in Sect. 3.3.2. The sequences of recognized gestures are delayed with respect to the sequence of target gestures (Fig. 4.20a). Both signals were time-synchronized (Fig. 4.20b) by calculating the delay based on the cross-correlation (Sect. 3.4.2). The average time delay (T_d) used for the synchronization, which depends on the user response time (T_R) and the motion-selection time (MST), was 0.88 ± 0.14 s.

The user performance was evaluated by using lock-step distance metric (Sect. 3.4.2). The time series related to hand gestures are coded considering the following order: "Open < Rest < Closed" ("Open" $= -1$, "Rest" $= 0$ and "Close"

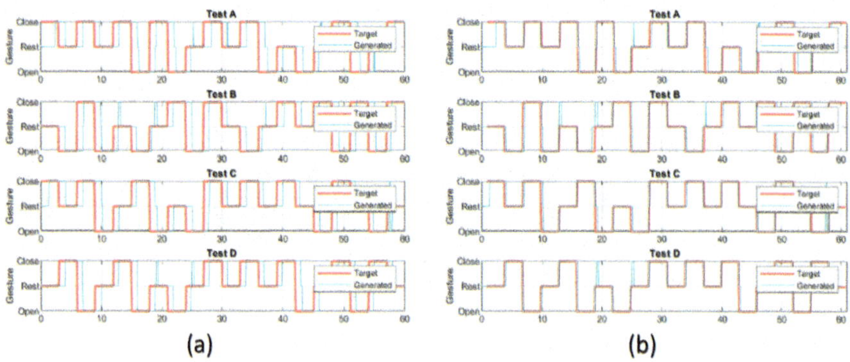

Fig. 4.20 Target and recognized sequence of gestures of one subject **a** raw data; **b** time-synchronized data [3]

= 1). As an error between open and closed gesture is more serious than an error between either of them or rest gesture, the L2 norm is used since the quadratic cost penalizes this type of error more than the L1-norm. Hence, the L2 distance between the target gesture and the synchronized gesture time series is computed as a similarity measurement.

Table 4.3 shows the results of the multifactorial additive ANOVA, which was performed to evaluate the influence of three independent variables (type of test, test order and individual) on the L2 distances.

No statistically significant differences were found in the order in which the tests were performed ($F(3) = 1.241$, $p = 0.3054$). Additionally, statistically significant differences were found in the average of L2 by type of test performed ($F(3) = 4.028$, $p = 0.0124$). A Duncan post-hoc test (Table 4.4) revealed significant pairwise differences between performance in tests A and C ($p = 0.0497$) and between performance in tests B and C ($p = 0.0412$).

The distribution of the L2 distances according to the type of performed test is shown in Fig. 4.21. L2 distances were 3.39 ± 0.70, 3.43 ± 0.75, 2.89 ± 0.71, 3.17 ± 0.73 for test A, B, C and D, respectively. Finally, homoscedasticity was verified.

An important feature of the wearable and multimodal embedded system presented in Sect. 4.1.2, apart from its notable computational capabilities and high versatility, is its power consumption. It has a battery life of 5 h and maximum power usage of around 1250 mW. To perform a power consumption analysis, the system is divided into three components: the sensor modules, TMS320F28069M MCU, and CC2650 MCU.

Regarding the sensor modules, when active and set to the defined specifications, the EMG, IMU, ECG, GSR, and TEMP modules consume 40.6 mW, 5.6 mW, 0.6 mW,

Table 4.3 Results of the multifactorial additive ANOVA assessing the impact of the three independent variables (type of test, test order and individual) on the L2 distances

		Df	Sum sq.	Mean sq.	F value	Pr (> F)
Test	*	3	3.366	1.1221	4.028	0.0124
Order		3	1.037	0.3456	1.241	0.3054
Individual	***	17	20.933	1.2313	4.420	$2.43\ e^{-05}$
Residuals		48	13.373	0.2786		

***Denotes significance at the (< 0.001) level and *at the (<0.5) level

Table 4.4 Results of the Duncan's multiple range test showing pairwise comparisons and significant differences among the variables

	Test A	Test B	Test C
Test B	0.8775	–	–
Test C	0.0497*	0.0412*	–
Test D	0.3557	0.3121	0.2451

*Denotes significance at the (< 0.5) level

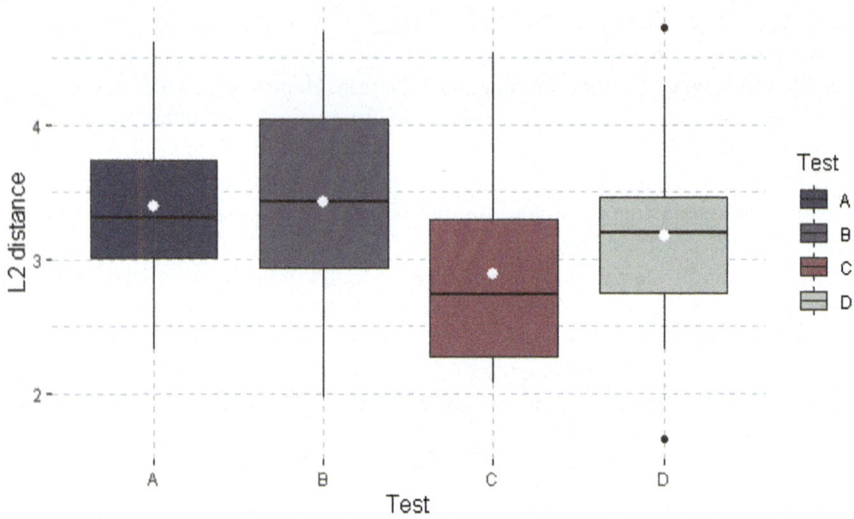

Fig. 4.21 L2 distances for the four performed tests [3]

3.3 mW, and 5.6 mW, respectively. In sleep mode, these modules consume 9.6 mW, 0.02 mW, 0.001 mW, 2.6 mW, and 0.01 mW, respectively.

Additionally, the power consumption of the TMS320F28069M MCU was analyzed for the two presented rehabilitation scenarios: EMG&IMU-based control for upper-limb VR-based exergame (Sect. 4.2.2), and the AAN control for a wrist rehabilitation robot (Sect. 4.2.3). Figure 4.22 details the average power consumption of each peripheral for each application. During data acquisition from the sensors, the DMA is triggered to transfer the data to the internal memory, while the CLA module independently processes the signal. The DMA's energy consumption is proportional to the number of active channels and the amount of data and transfer rate, while the CLA energy consumption depends on the algorithm's complexity.

The system's power consumption is heavily influenced by the application and MCU configuration, with the energy consumption of the MCU being highly variable based on the computational load. When the computation is over, the MCU can be programmed to enter an IDLE mode, which reduces power consumption from 272.3 to 82.5 mW. Motion recognition control has a power consumption of 203 mW, while adaptive control has a power consumption of 249 mW.

The CC2650 module's power consumption is primarily due to BLE communication, with the average power consumption being proportional to the amount of data transmitted. When streaming raw data from all five sensor modules at 1 kHz, the maximum power consumption for BLE communication can reach 32.7 mW. In the proposed scenarios, the CC2650 MCU only receives information through the i2C interface and sends it to a central node using BLE, with no additional modules or data processing being carried out, which would increase the module's power consumption.

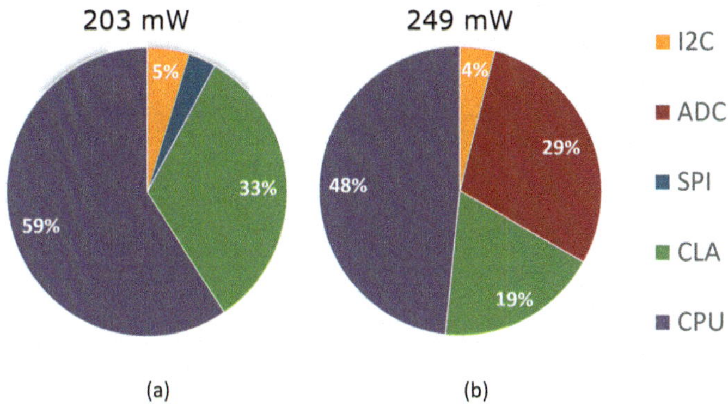

Fig. 4.22 Breakdown of power consumption in the two scenarios: **a** EMG&IMU-based control, and **b** AAN control [2]

References

1. Cisnal A, Perez-Turiel J, Fraile JC, Sierra D, De La Fuente E (2021) RobHand: a hand exoskeleton with real-time EMG-driven embedded control. Quantifying hand gesture recognition delays for bilateral rehabilitation. IEEE Access 9:137809–137823. https://doi.org/10.1109/ACCESS.2021.3118281
2. Cisnal A, Antolínez D, Turiel JP, Carlos Fraile J, De La Fuente E (2023a) A versatile embedded platform for implementation of biocooperative control in upper-limb neuromotor rehabilitation scenarios. IEEE Access 11:35726–35736. https://doi.org/10.1109/ACCESS.2023.3265898
3. Cisnal A, Gordaliza P, Pérez Turiel J, Fraile JC (2023) Interaction with a hand rehabilitation exoskeleton in EMG-driven bilateral therapy: influence of visual biofeedback on the users' performance. Sensors 23:2048. https://doi.org/10.3390/s23042048
4. Cisnal A, Sierra D, Turiel JP, Fraile JC (2020) An embedded implementation of EMG-driven control for assisted bilateral therapy. In: Torricelli D, Akay M, Pons JL (eds) Converging clinical and engineering research on neurorehabilitation IV. ICNR 2020. Biosystems & biorobotics. Springer, pp 817–821. https://doi.org/10.1007/978-3-030-70316-5_130
5. Open Source Sensor Fusion (2015). (WWW Document). URL https://github.com/memsindustrygroup/Open-Source-Sensor-Fusion
6. MIhelj M, Podobnik J (2012) Haptics for virtual reality and teleoperation. J Chem Inf Model. https://doi.org/10.1007/978-94-007-5718-9
7. Mihelj M, Novak D, Munih M (2009) Emotion-aware system for upper extremity rehabilitation. In: 2009 Virtual rehabilitation international conference, VR 2009, pp 160–165. https://doi.org/10.1109/ICVR.2009.5174225
8. Mandryk RL, Atkins MS (2007) A fuzzy physiological approach for continuously modeling emotion during interaction with play technologies. Int J Hum Comput Stud 65:329–347. https://doi.org/10.1016/j.ijhcs.2006.11.011
9. Guerrero CR, Fraile Marinero JC, Turiel JP, Muñoz V (2013) Using "human state aware" robots to enhance physical human-robot interaction in a cooperative scenario. Comput Methods Programs Biomed 112:250–259. https://doi.org/10.1016/j.cmpb.2013.02.003
10. Pan J, Tompkins WJ (1985) A real-time QRS detection algorithm. IEEE Trans Biomed Eng BME 32:230–236. https://doi.org/10.1109/TBME.1985.325532
11. Muñoz JE, Gouveia ER, Cameirão MS, Badia SBI (2018) Physiolab—a multivariate physiological computing toolbox for ECG, EMG and EDA signals: a case of study of cardiorespiratory fitness assessment in the elderly population. Multimed Tools Appl 77:11511–11546. https://doi.org/10.1007/s11042-017-5069-z

Chapter 5
Discussion

In this Doctoral Thesis, the main aspects that limit the use of biocooperative systems have been addressed: hardware and reliability. First, a real-time embedded system is developed for the acquisition of EMG signals. This system is integrated into a hand robotic exoskeleton, enabling EMG-driven bilateral assistive therapies. The evaluation focuses on assessing the impact of the real-time embedded system on the performance of the developed EMG-driven rehabilitation platform. The accuracy and responsiveness of the system were determined; these are essential parameters for assessing the quality of the human–robot interaction. The proposed EMG-driven rehabilitation system is further investigated by incorporating EMG-based visual feedback. The influence of this visual biofeedback on the user's performance during the execution of bilateral therapies is evaluated. The findings suggest that this type of feedback could help patients increase their control over the movement of the robotic platform. Lastly, a wearable real-time embedded system for multimodal signal acquisition was developed and tested by implementing two biocooperative control strategies. This versatile, low-cost system provides reliable signals for implementing real-time biocooperative controls across a wide range of neuromotor rehabilitation applications.

In this chapter, the findings are discussed in accordance with the hypothesis and results presented in previous chapters. Each finding is given its own dedicated section, resulting in three distinct sections. Within these sections, a comprehensive analysis of the research findings and their implications in the field of rehabilitation robotics is provided.

A. Cisnal de la Rica, *Development of Biocooperative Control Strategies for Neuromotor Rehabilitation Robotic Platforms*, Springer Theses,
https://doi.org/10.1007/978-3-032-02485-5_5

5.1 EMG-Driven Hand Rehabilitation Robot

A systematic review according to PRISMA (Reporting Items for Systematic Reviews and Meta-Analyses) was conducted to compare the performance in terms of accuracy and latency of the proposed EMG-driven robotic rehabilitation platform with other similar works. Articles were searched in various databases (IEEE Explorer, Web of Science and PubMed) using specific keywords and inclusion/exclusion criteria were searched to answer the question "What is the latency time of EMG-driven hand rehabilitation robots?". Nine articles met the criteria; however, only two reported latency times (Table 5.1).

The overall accuracy of the gesture recognition module was 0.97, and misclassifications during the gesture were small, which means that the error is not perceived in the exoskeleton in most cases. The accuracy results are comparable to other control systems proposed in the literature, varying from 77.9 to 98.7% (Table 5.1).

Regarding the responsiveness analysis, [4] conducted a temporal analysis of a hand exoskeleton controlled by EMG signals. However, the five time parameters provided are not readily comparable with our findings. Additionally, they were inconclusive regarding the delay experienced by the user, as they were primarily influenced by the mechanical structure of the exoskeleton and the waiting time for commands. In [5], the "control speed" was measured as the number of actions correctly performed in one minute and provided an average "time per action" of 1.4 s, which is comparable to the MCT (the MCT of the proposed control in the RobHand is 1.90 ± 1.65 s). However, the "time per action" value alone does not provide enough information

Table. 5.1 Published data regarding the accuracy and temporal information of EMG-driven robots for hand rehabilitation

Ref.	Control type	Accuracy	Delay[2]
[1]	Point-in-polygon (PIP) (3 predefined gestures)	Accuracy: 0.944	–
[2]	EMG-driven (neural network to determine force)	Force error: 20.7%	–
[3]	Threshold algorithm	–	–
[4]	Linear Bayes classifier (6 predefined gestures)	Accuracy[1]: $98.1 \pm 4.9\%$	Yes
[5]	Classification algorithm (6 predefines gestures)	Accuracy: 86.38%	Yes
[6]	EMG-based intent inference method	–	–
[7]	Neural network	–	–
[8]	Forest classifier (3 predefined gestures)	Accuracy: 77.9 - 85.2%	–
[9]	Neural Network (4 predefined gestures)	Accuracy: $98.7 \pm 0.53\%$	–

[1] Accuracy for neurologically intact subjects
[2] Whether delay analysis of the system is carried out

about the lag, as the computational time could be very large while the actuators could be very fast, resulting in a low "time per action."

Due to the limited results, the systematic review was expanded to look for embedded systems for EMG gesture recognition integrated in hand rehabilitation robots. It yielded some studies that specify the computational time of the classification algorithm, ranging from 0.58 to 2.8 ms [10–12], but comparisons are not possible due to the lack of information on EMG acquisition and data preprocessing. More information on control latency was available in studies on hand prostheses based on EMG, with reported times ranging from 100 to 600 ms [13–15], but no reliable comparison can be made due to inaccurate definitions and uncertain time periods.

In summary, the presented system is characterized by its low-cost embedded EMG acquisition system, which has enabled the implementation of a real-time EMG-driven control for performing bilateral therapies with a hand robotic exoskeleton. The distinctive feature of the designed 2-channel EMG acquisition device is its cost, estimated at approximately 30€ for low-scale production. The key advantage of implementing EMG-based control within the real-time embedded system lies in the reduction of latency, while maintaining high accuracy in gesture recognition.

5.2 Influence of Visual Biofeedback on the Users' Performance

Significant statistical differences in subject performance were observed during execution of EMG-driven bilateral therapies based on the type of feedback provided. Specifically, users' performance significantly improved when only the EMG-based visual feedback (test C) was present compared to the presence of only kinesthetic feedback (test B) and both feedbacks (tests A). Therefore, it can be concluded that the visual biofeedback enhanced the user motor control. It provided users with a simple way to monitor and regulate their EMG activation levels with respect to the previously predefined activation thresholds, and thus enabling them to have better self-control the movement of the hand exoskeleton.

The biofeedback is provided earlier in time than the kinesthetic feedback and therefore, the user has a longer reaction time to self-regulate their EMG activation levels. In fact, the user can directly adjust the force being exerted at the moment by influencing the position controller input through real-time visual EMG feedback. On the other hand, kinesthetic feedback requires the user to first feel the movement performed by the exoskeleton's actuators before modulating the exerted force, resulting in a delayed force modulation (Fig. 5.1). Consideration should be given to the electromechanical characteristics of the actuators, including their low dynamic response and limited speed due to the rehabilitation application. Note that relevant time delays for the human–robot interaction have been preciously determined, MST, MOT and MCT.

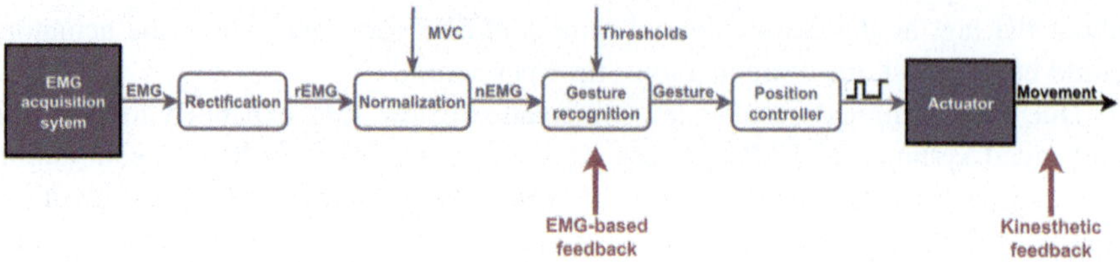

Fig. 5.1 The control loop used for the threshold EMG-driven control of the RobHand specifying the origin of each feedback [16]

On the other hand, the fact that EMG-biofeedback results in significantly higher performance than in the presence of both feedbacks may be due to the fact that the kinesthetic feedback is more straightforward than the visual biofeedback and hence, users do not need to consciously pay attention to it.

In summary, in the presence of visual feedback, the user adjusts his/her force based on the data provided by the gesture recognition module (nEMG signals and detected gesture) and anticipates the response of the exoskeleton. On the other hand, with kinesthetic feedback, the user adjusts his/her force after the exoskeleton has performed the action. If the user notices that the movement performed by the exoskeleton does not align with their intention, the user can correct it by modulating their muscle activity, but this process takes longer than if they had corrected it based on real-time EMG visual feedback.

5.3 Wearable Embedded Multimodal Acquisition System

The research introduced a cost-effective wearable embedded multimodal platform designed for the implementation of biocooperative control in the field of neuromotor rehabilitation. A significant observation from the examination of related studies in biocooperative control is that most of them incorporated only a restricted set of sensors and relied on expensive and often bulky commercial systems, lacking the necessary processing capabilities.

One approach to reduce the bulkiness of acquisition systems is the integration of sensors directly into the system. For instance, a previous study proposed integrating multiple sensors, including ECG, GSR, SKT, PPG, and force sensors, into the handle of an end-effector rehabilitation [17]. Similarly, in other fields, researchers have embedded ECG, GSR, accelerometers, and force sensors onto wheelchairs [18] or integrated direct contact sensors onto steering wheels to measure ECG, GSR, SpO2 levels, and SKT signals [19]. However, it should be noted that while this integration approach reduces bulkiness, it may also limit flexibility.

Recent research emphasizes the need for wearable sensing devices that are small, low-cost, and possess high computational power to detect human physical activity and emotions through multimodal fusion strategies. In order to promote portability and comfort, the number of sensors used must be chosen carefully, ensuring that the system is fast, energy-efficient, and convenient. Additionally, a smaller set of sensors placed in optimal locations can increase user acceptance [20].

Several studies have attempted to create affordable wearable platforms. For instance, some works have focused on the development of integrated EMG sensors [11, 21, 22], while others have specifically developed ECG sensors [23, 24], or solely utilized IMU sensors [25]. In contrast, other works have integrated more than one sensor into the platform, such as ECG and GSR [26], IMU and ECG [27] and GSR, SKT and IMU [28]. However, these studies have not sufficiently addressed the need for versatility in terms of the number of sensors required to implement biocooperative control systems.

The presented wearable, multimodal and low-cost solution overcome these limitations. It integrates five sensors (IMU, EMG, SKT, GSR, and ECG) and its highly configurable, thus offering a more comprehensive approach for biocooperative control strategies in the context of neurorehabilitation. The platform's high-efficiency real-time MCU provides ample processing capabilities and its high flexibility, concerning sensor diversity and wireless communication, allows the development of numerous rehabilitation applications.

Two potential applications using different signals and rehabilitation approaches were proposed. Arm motion tracking using IMU data, hand gesture recognition through EMG signals, HR detection from ECG signals, SCL and SCR extraction from GSR data and SKT monitoring were carried out. It was verified that the system's low cost does not compromise the quality of the signals, that is reliable enough for the proposed scenarios and good performance is expected in other related applications. Additionally, the system power consumption for the two scenarios were analyzed and showed that they are within the energy constraints of the system. It can record and process real-time multimodal information for at least 5 h.

The wearable system was designed by balancing complexity, price, and performance, taking into account parameters such as volume, flexibility, energy consumption, onboard processing, and signal quality. Due to its low cost, compact size, and comfort, the platform shows great potential for rehabilitation applications. While it may not be appropriate for applications that demand high signal quality or a long-term battery life, such requirements are not typically expected in rehabilitation applications. Therefore, the presented platform is a promising advancement in the development of wearable technologies for neuromotor rehabilitation.

References

1. Secciani N, Topini A, Ridolfi A, Meli E, Allotta B (2020) A novel point-in-polygon-based sEMG classifier for hand exoskeleton systems. IEEE Trans Neural Syst Rehabil Eng 28:3158–3166. https://doi.org/10.1109/TNSRE.2020.3044113
2. Leonardis D, Barsotti M, Loconsole C, Solazzi M, Troncossi M, Mazzotti C, Castelli VP, Procopio C, Lamola G, Chisari C, Bergamasco M, Frisoli A (2015) An EMG-controlled robotic

hand exoskeleton for bilateral rehabilitation. IEEE Trans Haptics 8:140–151. https://doi.org/10.1109/TOH.2015.2417570
3. Ben IA, Bouteraa Y, Rekik C (2017) Design and development of 3D printed myoelectric robotic exoskeleton for hand rehabilitation. Int J Smart Sens Intell Syst 10:341–366. https://doi.org/10.21307/ijssis-2017-215
4. Lu Z, Chen X, Zhang X, Tong KY, Zhou P (2017) Real-time control of an exoskeleton hand robot with myoelectric pattern recognition. Int J Neural Syst 27:1–11. https://doi.org/10.1142/S0129065717500095
5. Zhang J, Wang B, Zhang C, Xiao Y, Wang MY (2019) An EEG/EMG/EOG-based multimodal human-machine interface to real-time control of a soft robot hand. Front Neurorobot 13:1–13. https://doi.org/10.3389/fnbot.2019.00007
6. Park S, Fraser M, Weber LM, Meeker C, Bishop L, Geller D, Stein J, Ciocarlie M (2020) User-driven functional movement training with a wearable hand robot after stroke. IEEE Trans Neural Syst Rehabil Eng 28:2265–2275. https://doi.org/10.1109/TNSRE.2020.3021691
7. Burns MK, Pei D, Vinjamuri R (2019) Myoelectric control of a soft hand exoskeleton using kinematic synergies. IEEE Trans Biomed Circuits Syst 13:1351–1361. https://doi.org/10.1109/TBCAS.2019.2950145
8. Park S, Meeker C, Weber LM, Bishop L, Stein J, Ciocarlie M (2018) Multimodal sensing and interaction for a robotic hand orthosis. IEEE Robot Autom Lett 4:315–322. https://doi.org/10.1109/LRA.2018.2890199
9. Chen Y, Yang Z, Wen Y (2021) A soft exoskeleton glove for hand bilateral training via surface EMG. Sensors 21:1–18. https://doi.org/10.3390/s21020578
10. Benatti S, Milosevic B, Casamassima F, Schonle P, Bunjaku P, Fateh S, Huang Q, Benini L (2014) EMG-based hand gesture recognition with flexible analog front end. In: 2014 IEEE biomedical circuits and systems conference (BioCAS) proceedings. IEEEE, Lausanne, Switzerland, pp 57–60. https://doi.org/10.1109/BioCAS.2014.6981644
11. Benatti S, Casamassima F, Milosevic B, Farella E, Schönle P, Fateh S, Burger T, Huang Q, Benini L (2015) A Versatile embedded platform for EMG acquisition and gesture recognition. IEEE Trans Biomed Circuits Syst 9:620–630. https://doi.org/10.1109/TBCAS.2015.2476555
12. Liu J, Zhang F, Huang HH (2014) An open and configurable embedded system for EMG pattern recognition implementation for artificial arms. In: 2014 36th Annual international conference of the IEEE engineering in medicine and biology society (EMBC). IEEE, Chicago, IL, USA, pp 4095–4098. https://doi.org/10.1109/EMBC.2014.6944524
13. Chu JU, Moon I, Mun MS (2006) A real-time EMG pattern recognition system based on linear-nonlinear feature projection for a multifunction myoelectric hand. IEEE Trans Biomed Eng 53:2232–2239. https://doi.org/10.1109/TBME.2006.883695
14. Ryser F, Bützer T, Held JP, Lambercy O, Gassert R (2017) Fully embedded myoelectric control for a wearable robotic hand orthosis. In: 2017 International conference on rehabilitation robotics (ICORR). IEEE, London, UK, pp 615–621. https://doi.org/10.1109/ICORR.2017.8009316
15. Tam S, Boukadoum M, Campeau-Lecours A, Gosselin B (2020) A fully embedded adaptive real-time hand gesture classifier leveraging HD-sEMG and deep learning. IEEE Trans Biomed Circuits Syst 14:232–243. https://doi.org/10.1109/TBCAS.2019.2955641
16. Cisnal A, Gordaliza P, Pérez Turiel J, Fraile JC (2023) Interaction with a hand rehabilitation exoskeleton in emg-driven bilateral therapy: influence of visual biofeedback on the users' performance. Sensors 23:2048. https://doi.org/10.3390/s23042048
17. Jakopin B, Mihelj M, Munih M (2017) An unobtrusive measurement method for assessing physiological response in physical human-robot interaction. IEEE Trans Hum Mach Syst 47:474–485. https://doi.org/10.1109/THMS.2017.2681434
18. Postolache O, Viegas V, Dias Pereira JM, Vinhas D, Girão PS, Postolache G (2014) Toward developing a smart wheelchair for user physiological stress and physical activity monitoring. In: 2014 IEEE international symposium on medical measurements and applications (MeMeA). IEEE, Lisboa, Portugal, pp 1–6. https://doi.org/10.1109/MeMeA.2014.6860097
19. Heuer S, Chamadiya B, Gharbi A, Kunze C, Wagner M (2010) Unobtrusive in-vehicle biosignal instrumentation for advanced driver assistance and active safety. In: Proceedings of 2010 IEEE

EMBS conference on biomedical engineering and sciences (IECBES). IEEE, Kuala Lumpur, Malaysia, pp 252–256. https://doi.org/10.1109/IECBES.2010.5742238
20. Qiu S, Zhao H, Jiang N, Wang Z, Liu L, An Y, Zhao H, Miao X, Liu R, Fortino G (2022) Multi-sensor information fusion based on machine learning for real applications in human activity recognition: State-of-the-art and research challenges. Inform Fusion 80:241–265. https://doi.org/10.1016/j.inffus.2021.11.006
21. Brunelli D, Farella E, Giovanelli D, Milosevic B, Minakov I (2016) Design considerations for wireless acquisition of multichannel sEMG signals in prosthetic hand control. IEEE Sens J 16(23):8338–8347. https://doi.org/10.1109/JSEN.2016.2596712
22. Örücü S, Selek M (2019) Design and validation of multichannel wireless wearable SEMG system for real-time training performance monitoring. J Healthcare Eng 2019:1–15. https://doi.org/10.1155/2019/4580645
23. Athavipach C, Pan-Ngum S, Israsena P (2019) A wearable in-ear EEG device for emotion monitoring. Sensors 19(4014):1–16. https://doi.org/10.3390/s19184014
24. Nguyen A, Alqurashi R, Raghebi Z, Banaei-Kashani F, Halbower AC, Vu T, Lane N, Zhou X (2017) LIBS: A lightweight and inexpensive in-ear sensing system for automatic whole-night sleep stage monitoring. GetMobile: Mobile Comput Commun 21(3):31–34. https://doi.org/10.1145/3161587.3161596
25. Marta G, Pedrocchi A, Ferrante S, Costa A, Bortolotti D, Sorti S, Vitale, F, Bocciolone M, Braghin F, Masiero S. (2020) Wearable biofeedback suit to promote and monitor aquatic exercises: a feasibility study. IEEE Trans Instrum Meas 69(4):1219–1231. https://doi.org/10.1109/TIM.2019.2911756
26. Villar BF, de la Rica AC, Vargas MM, Turiel JP, Juan Carlos Fraile M. (2021) A low Cost IoT enabled device for the monitoring, recording and communication of physiological signals. In: Proceedings of the14th International Joint Conference on Biomedical Engineering Systems and Technologies (BIOSTEC), 1:BIODEVICES:135–143. https://doi.org/10.5220/0010302901350143
27. D'Mello Y, Skoric J, Xu S, Roche PJR, Lortie M, Gagnon S, Plant DV (2019) Real-time cardiac beat detection and heart rate monitoring from combined seismocardiography and gyrocardiography. Sensors (Switzerland) 19(16). https://doi.org/10.3390/s19163472
28. Huan J, Bernstein JS, DIfuntorum P, Masna NVR, Gravenstein N, Bhunia S, Mandal S (2022) A wearable skin temperature monitoring system for early detection of infections. IEEE Sens J 22(2):1670–1679. https://doi.org/10.1109/JSEN.2021.3131500

Chapter 6
Conclusions

In recent years, the development of affordable biocooperative rehabilitation systems for improving the QOL of people suffering from neuromotor disabilities has become a major challenge. Despite the growing interest of scientific literature in neuromotor rehabilitation robotics, limitations in current systems have made them accessible only to large academic medical centers and research laboratories. The excessive cost of physiological recording systems limits the accessibility of biocooperative controls for neuromotor rehabilitation even further.

Current research focuses on designing innovative controls aiming to achieve a more natural HRI that enhances patient motivation towards rehabilitation. However, these studies overlook the potential utilization of biocooperative systems in clinical practice and focus merely on academic purposes. For instance, numerous EMG-based robotic control strategies have been suggested in the literature. These approaches can detect numerous hand gestures with high accuracy. However, they require robots with multiple degrees of freedom and entail substantial computational expenses, rendering them infeasible and unreliable for practical applications. Moreover, these techniques rely on numerous EMG signals, which call for several electrodes to be fixed onto the patient, thereby adding to the already time-intensive clinical regimen of physical therapy personnel. Reducing the number of electrodes can lower costs, decrease power consumption, and improve user comfort.

The present Doctoral Thesis is focused on providing affordable technology and developing control strategies to enable real use of biocooperative systems by motor-disabled people. In this chapter, the main contributions and conclusions of the articles included in this compendium of publications are indicated in Sect. 6.1 and Sect. 6.2, respectively. Section 6.3 outlines future research directions related to work.

A. Cisnal de la Rica, *Development of Biocooperative Control Strategies for Neuromotor Rehabilitation Robotic Platforms*, Springer Theses,
https://doi.org/10.1007/978-3-032-02485-5_6

6.1 Contributions

In this section, the contributions of three articles that address the use of technology in neuromotor rehabilitation are highlighted. The first article presents a hand exoskeleton that supports EMG-driven assisted rehabilitation by using a custom-made low-cost EMG real-time embedded solution. The second article evaluates the influence of EMG-based visual biofeedback on user performance when performing EMG-driven bilateral exercises with the robotic hand exoskeleton. Finally, the third article introduces a low-cost and wearable embedded system that integrates the most commonly used sensors in neuromotor rehabilitation and hence, enables the development of real-time biocooperative controls for a wide range of applications in this field. These articles showcase the potential of technology in promoting motor recovery and improving rehabilitation outcomes in patients with neuromotor impairments. The main contributions and potential impact in the field of neuromotor rehabilitation provided by the results of this compendium of publications are the following:

I. Design and development of a low-cost EMG acquisition system for real-time EMG-driven therapies. Integration of the system in the Robhand rehabilitation platform, including additional electronic circuitry [1].
II. Development of a non-pattern recognition-based EMG-driven control for bilateral robotic hand rehabilitation. Integration of the control in the real-time embedded platform [1].
III. Performance evaluation of the EMG-driven exoskeleton in terms of accuracy and latency and comparison with previous works. It was detected that the HRI evaluation of EMG-based robotic rehabilitation systems in the current literature is inadequate. While studies have focused on providing accuracy metrics to evaluate gesture recognition performance, they have failed to provide latency metrics, which are essential for achieving optimal HRI [1].
IV. Design and implementation of an EMG-based visual feedback. Although the usefulness of EMG feedback in neuromotor rehabilitation applications has been demonstrated, to the best of our knowledge, this was the first time that EMG feedback combined with rehabilitation robots have been investigated [2].
V. Evaluation of the performance of the users when providing EMG-based visual feedback during EMG-driven bilateral therapies. We found that EMG feedback led to significant improvements in performance [2].
VI. Design and development of a low-cost, wearable, embedded and multimodal acquisition system for the implementation of biocooperative systems in neuromotor rehabilitation. The feasibility of the system and its power consumption was tested [3].
VII. Design and development of two biocooperative control strategies for neuromotor rehabilitation using the proposed embedded multimodal acquisition system. An EMG&IMU-based control using VR-based therapies and AAN control using a wrist rehabilitation robot [3].

6.2 Main conclusions

The development and evaluation of the proposed low-cost biocooperative rehabilitation systems have shown promising results [1, 3]. These systems have proven the feasibility of implementing biocooperative controls without the need for expensive commercial systems that are typically bulky and lack processing capabilities. The advantages associated with these embedded solutions for physiological signal acquisition, in addition to their low cost and versatility, include enabling the development of real-time biocooperative controls with reduced latency while maintaining high accuracy. These characteristics make embedded acquisition solutions an attractive option for the development of biocooperative control systems that are accessible and affordable for rehabilitation in clinical settings. Furthermore, the use of these systems can also pave the way for the development of new applications in the fields of human-robot interaction and assistive robotics.

It has been shown that the incorporation of EMG-based visual feedback can significantly improve the performance of individuals undergoing EMG-driven assisted therapies. By providing real-time biofeedback, the subjects were able to monitor and modulate their EMG responses, resulting in better control of the exerted force. Thus, it can be inferred that EMG-based visual feedback has the potential to facilitate the rehabilitation learning process, as it helps users develop a better understanding of how to self-regulate their muscle activations. This could ultimately enhance patient motivation and contribute to better motor recovery outcomes [2].

As a last point, there is a need for greater emphasis on responsiveness and latency times in EMG-driven rehabilitation systems [1]. Previous studies have primarily focused on the accuracy of gesture classification, neglecting important information regarding system latency. While some studies have addressed system latency, the measured values are often imprecisely defined. It is crucial to establish clear and standardized definitions of latency times for rehabilitation robotics to ensure homogeneity in this field of research. By doing so, we can ensure that the development of EMG-based rehabilitation systems is optimized for the benefit of patients.

6.3 Future research lines

Research on neurorehabilitation using biocooperative systems has shown promising results in recent years, highlighting the potential of these technologies to improve the quality of life of patients undergoing neurorehabilitation. Several future research questions can be derived from this investigation.

Firstly, it is important to focus on people with neurological disabilities. While this research has shown positive results in healthy individuals, it is important to explore the potential benefits for motor recovery in individuals with neurological impairments. By doing so, it may be possible to identify specific adaptations or modifications that can further enhance its effectiveness for this population. Additionally, it would be convenient to develop a unified protocol for the evaluation of the human-robot interface by measuring latency times, which would allow comparisons between different rehabilitation robots.

The use of EMG-based visual feedback is another area where future research can be directed. As technology has proven effective in healthy individuals through this study, it is crucial to investigate its potential advantages for those with neurological disorders. Additionally, the use of more complex visualizations of EMG-based feedback, such as in combination with other virtual reality objects, should be further explored to determine if it has a positive impact on user performance.

Finally, future research should focus on implementing more advanced biocooperative controls using the proposed multimodal embedded platform, which could provide further insights into the capabilities of the system. Additionally, the reduction of electronic size and energy consumption will improve its wearability, resulting in better user acceptance. Therefore, researchers should work towards developing energy-efficient and compact systems that could be easily integrated into wearable devices, making them more comfortable for the user to wear. This would not only improve user acceptance but also enable remote monitoring of patients' progress, leading to more personalized treatment plans.

Overall, the potential benefits of biocooperative systems in neurorehabilitation are clear, and future research should focus on refining these systems to better serve the needs of individuals with neurological impairments. Continued research in this area has the potential to make a significant impact on the quality of life for those undergoing neurorehabilitation.

References

1. Cisnal A, Perez-Turiel J, Fraile JC, Sierra D, De La Fuente E (2021) RobHand: a hand exoskeleton with real-time EMG-driven embedded control. Quantifying hand gesture recognition delays for bilateral rehabilitation. IEEE Access 9:137809–137823. https://doi.org/10.1109/ACCESS.2021.3118281
2. Cisnal A, Gordaliza P, Pérez Turiel J, Fraile JC (2023b) Interaction with a hand rehabilitation exoskeleton in EMG-driven bilateral therapy: influence of visual biofeedback on the users' performance. Sensors 23:2048. https://doi.org/10.3390/s23042048
3. Cisnal A, Antolínez D, Turiel JP, Carlos Fraile J, De La Fuente E (2023a) A versatile embedded platform for implementation of biocooperative control in upper-limb neuromotor rehabilitation scenarios. IEEE Access 11:35726–35736. https://doi.org/10.1109/ACCESS.2023.3265898

About the Author

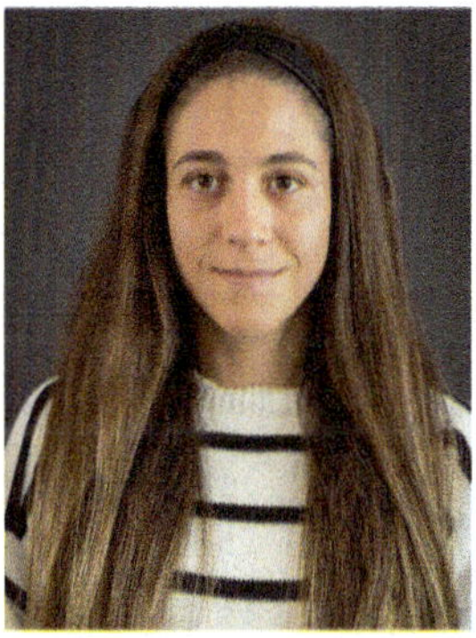

Ana Cisnal de la Rica earned her B.S. degree in Industrial Electronics and Automation Engineering (2017) and M.S. in Industrial Engineering (2019) from the University of Valladolid, Spain. In 2023, she completed her Ph.D. in Industrial Engineering at the same university with the highest honors (*Cum Laude*) and an international distinction.

During her Bachelor's program, she gained research experience at the Health Division of CARTIF (Valladolid, Spain), contributing to the development of the E2REBOT robotic platform for upper-limb neurorehabilitation, and at the Fraunhofer Institute for Biomedical Engineering (Sankt Ingbert, Germany), where she designed a measurement system for electrode characterization in the context of functional electrostimulation.

Since 2017, Ana has been affiliated with the Medical Robotics Research Group at the Advanced Production Technologies Institute (ITAP), University of Valladolid, where her work centers on neuromotor rehabilitation and surgical robotics. She has also collaborated with industry, developing wearable sensing solutions for livestock monitoring at IDIMAS Gestión SLNE. From 2020 to 2023, she served as an assistant professor at the University of Valladolid's Department of Systems Engineering

A. Cisnal de la Rica, *Development of Biocooperative Control Strategies for Neuromotor Rehabilitation Robotic Platforms*, Springer Theses,
https://doi.org/10.1007/978-3-032-02485-5

and Automation, and in 2024 joined the Universidad Europea Miguel de Cervantes (Valladolid) as an assistant professor.

She expanded her research internationally as a visiting Ph.D. student and postdoctoral researcher at the Spinal Cord Artificial Intelligence (SCAI) Lab, working on personalized healthcare for individuals with spinal cord injuries. In 2024, she joined the Robotics Lab as a visiting postdoctoral researcher, focusing on control strategies for therapeutic robotic massage. As a result of her research activity, she has published 18 articles in international journals indexed in the JCR and presented 16 papers at conferences, including 10 national and 6 international events.

The manufacturer's authorised representative in the EU is Springer Nature Customer Service Centre GmbH, Europaplatz 3, 69115 Heidelberg, Germany. If you have any concerns regarding our products, please contact ProductSafety@springernature.com

Printed and bound by CPI Group (UK) Ltd, Croydon, CR0 4YY
30/09/2026
02225471-0004